Cambridge Physics in the Thirties

Cambridge Physics in the Thirties

edited and introduced by

John Hendry

University of London

Adam Hilger Ltd, Bristol

This compilation © John Hendry 1984

British Library Cataloguing in Publication Data

Cambridge Physics in the Thirties
  1. Cavendish Laboratory—History
  I. Hendry, John
  507′.20426′59 Q183.E72C

  ISBN 0-85274-761-6

Consultant Editor: **Professor A J Meadows**,
University of Leicester

Published by Adam Hilger Ltd, Techno House,
Redcliffe Way, Bristol BS1 6NX

Phototypeset by Input Typesetting Ltd, London SW19 8DR and printed
in Great Britain by J W Arrowsmith Ltd, Bristol

Contents

Preface

There have been many periods of excitement in the history of experimental physics, but never has there been anything to compare with that described in this volume. Just half a century ago, between about 1930 and 1935, the Cavendish Laboratory at Cambridge provided the scene for an unprecedented sequence of breakthroughs and discoveries in the course of which nuclear physics as we know it was born.

Since then, many recollections of the Cavendish and especially of its director, Lord Rutherford, have been written and published. But as the fiftieth anniversary of the splitting of the atom and the discovery of the neutron and positron approached, it became clear that much still lay unrecorded. The first aim of this volume was to rectify that defect by publishing for the first time the recollections of some of the distinguished veterans of the Cavendish of the twenties and thirties. In addition, it was noted that many of the existing published recollections of the period were either little known or else not easily available, and it was therefore thought desirable to include a number of previously published essays written by those stars of the period who are now no longer with us. Finally, an attempt has also been made, through editorial direction and suggestions, and through the inclusion of introductory sections and full references, to create a volume that will not just be a collection of reminiscences, but will also serve as a useful introduction to the history of the events described.

The resulting compilation owes much to many people. In particular I should like as editor to record my appreciation of the tremendous helpfulness and cooperation of the contributors, some of whom have given far more to the volume than the contents page can indicate. I should also like to thank Margaret Gowing, Jack Meadows, Neville Goodman and Lady Constanza Blackett for their encouragement of the project at different stages. All responsibility for the content of the editorial sections and for the editing and contextual setting of the contributions is of course my own.

JOHN HENDRY

Acknowledgments

The articles by Norman Feather and James Chadwick were originally published in the *Proceedings of the Tenth International Congress in the History of Science, Ithaca, 1962* (Paris: Hermann, 1964), pp 135–45 and 159–62 respectively. They are reprinted here by kind permission of Michael Feather, Lady Eileen Chadwick, and Hermann and Company.

The article by Patrick Blackett is extracted from his Nobel Prize Lecture of 1948, published in *Nobel Lectures in Physics, 1942–1962* (Amsterdam: Elsevier, 1964), pp 97–108. It is reprinted here by kind permission of Lady Constanza Blackett and the North-Holland Publishing Company.

The article by John Cockcroft was originally published in the *Proceedings of the International Conference on Fast Neutron Physics, Rice University, 1963* (Houston, Texas: Rice University, 1964). It is reprinted here by kind permission of Lady Elizabeth Cockcroft and Rice University.

The article by Mark Oliphant is extracted from his book *Rutherford: Recollections of the Cambridge Days* (Amsterdam: Elsevier, 1972), pp 104–11. It is reprinted here by kind permission of the author and the North-Holland Publishing Company.

Contributors

Professor T E Allibone, CBE, FRS
Baron Blackett of Chelsea, FRS†
Baron Bowden of Chesterfield
Sir James Chadwick, FRS†
Sir John Cockcroft, FRS†
N A de Bruyne, FRS
Professor P I Dee, CBE, FRS†
Professor P A M Dirac, OM, FRS
W E Duncanson, OBE
Professor N Feather, FRS†
Professor M Goldhaber
Professor W B Lewis CBE, FRS
Sir Harrie Massey, FRS†
Sir Nevill Mott, FRS
Sir Mark Oliphant, AC, FRS
Sir Rudolf Peierls, FRS
Professor E T S Walton
C E Wynn-Williams†
Sir Alan Wilson, FRS

† Deceased

General Introduction

'It never rains but it pours.' So wrote Lord Rutherford in April 1932 to his great friend and colleague, the famous Danish physicist Niels Bohr.[1] Less than two months before that, James Chadwick had discovered the neutron. Now John Cockcroft and Ernest Walton had performed the first artificial disintegration of lithium, the splitting of the atom. Chadwick, Cockcroft and Walton were all members of Rutherford's team at the Cavendish Laboratory in Cambridge, and the achievements of that laboratory in 1932 and the ensuing years have since become legendary. Before the year was out Patrick Blackett and Giuseppe Occhialini had demonstrated the existence of yet another new particle, the positron. The following year Mark Oliphant and Rutherford himself observed the fusion of two deuterium or heavy hydrogen nuclei, with an accompanying release of nuclear energy. In 1934 Chadwick and Maurice Goldhaber demonstrated the nuclear photoelectric effect and derived the first accurate figure for the mass of the neutron. Of those involved in the achievements of 1932, Chadwick, Cockcroft and Walton all received Nobel Prizes for their work. Blackett's experiments of that year were also among those for which he received his Nobel Prize.

Most heads of department would have been delighted to play host to any one of the historic experiments conducted at Cambridge in the early 1930s. For such a string of successes to have come from a single laboratory is unparalleled in the history of physics. Fifty years later, this volume of recollections takes a look back at those golden years of Cambridge physics through the eyes of some of the physicists working there at the time, and asks what lay behind the achievements, what made the Cavendish Laboratory of that period so special.

In Part 1 we focus on the three famous experiments of 1932, those of Chadwick, Cockcroft and Walton, and Blackett and Occhialini. As well as starting a new era in physics these were also the culminating achievements of a series of investigations in laboratories throughout the world, stretching back into the 1920s. Chadwick's discovery emerged from research programmes that had been continuing not only at the Cavendish but also in Paris and Berlin. Cockcroft and Walton, in achieving the first artificial disintegration of an atomic nucleus, only narrowly beat rival teams working towards the same end with much more powerful apparatus in America. The work of Blackett and Occhialini brought technical advances and theoretical

1

developments made in Cambridge to bear in a field, the study of cosmic rays, that had been pioneered abroad. In the introduction to Part 1 we trace the internal scientific developments leading to the Cambridge experiments, and describe briefly what these experiments entailed. This is followed by some recollections, composed by Norman Feather and James Chadwick in the early 1960s, of the search for, and ultimate discovery of, the neutron. Feather worked very closely with Chadwick, and on the discovery of the neutron he turned immediately to the investigation of its properties, taking within weeks some beautiful cloud chamber photographs of the collisions between neutrons and nitrogen nuclei. In his recollections he describes the technical and scientific background to Chadwick's work, comparing the different approaches, attitudes and techniques adopted in the related programmes of Cambridge, Berlin and Paris, and offering his reflections as to how Chadwick came to enter this field personally and quickly come out on top. Chadwick himself then records his repeated attempts to observe neutrons, whose existence had been predicted by Rutherford, in the ten years preceding his final and somewhat unexpected success. Concluding the neutron story Philip Dee, who like Feather began investigating the neutron very soon after its discovery, recalls the course of his own investigations and how he, Rutherford and Chadwick reacted to them at the time.

Continuing the recollections of Part 1, Ernest Walton recalls some of the details of his pioneering work with Cockcroft, and sets these in the context of his own attitude to experimental physics. Patrick Blackett, in an extract from his Nobel Prize Lecture of 1948, describes his experiments with cloud chambers during the 1920s and early 1930s, and places his demonstration of the existence of the positron in the context of his wider research programme and of his general approach to experimental physics. Finally Paul Dirac, who in 1932 was the newly appointed Lucasian Professor of Mathematics at Cambridge, recalls his discussions with Blackett on the subject of the positron, whose existence he had predicted theoretically.

In Part 2 we turn from the famous experiments themselves to their context, to the Cavendish Laboratory and its members, to the teaching of physics in Cambridge, and to the research being conducted there both inside and outside the laboratory. We begin by introducing the leading members of the physics community in 1932 and noting the research on which they were engaged. In this introduction we also look briefly outside the Cavendish Laboratory, at the Cambridge mathematical physicists, at the lectures on offer to students, and at the general institutional context within which the physicists were working. By way of recollections we include some reminiscences written by John Cockcroft about twenty years ago of his own experiences and activities in the Cavendish during the 1920s and early 1930s. Norman de Bruyne, who studied and researched in the laboratory from 1923 to 1930, then offers his recollections of laboratory life, of the lectures he attended, and of the trials and tribulations of being a research student. W E Duncanson recalls his experience as one of Chadwick's research students during the exciting period from 1930–4, when he got to know the quiet Chadwick better than have many people. Finally Harrie Massey recalls the varied activities of the Cavendish during the 1931–2 session, as seen by one working in the laboratory but outside the main nuclear physics programme.

Of the many factors making up the context of the Cambridge *annus mirabilis* of 1932, several stand out as being of particular historical interest, and it is to three of these factors that Part 3 of this volume is devoted. The first concerns the relationships between theory and experiment, and between theoretical and experimental physicists. The reputation of the Cavendish is as a laboratory whose members were not greatly interested in advanced theoretical developments, and there is an element of truth in this. But there is another side to the story and in some ways, ways that were important for the discoveries of the 1930s, theory and experiment were very closely linked indeed. We therefore look at the theoretical backgrounds of some of the leading Cavendish physicists and at the role played by Ralph Fowler, who directed a theoretical group within the laboratory. We look generally at the theoretical work being pursued in Cambridge during the early 1930s, and in particular at the contacts between experimentalists and theoreticians, most notably through the Kapitza Club.

The second and third factors of interest concern the development of new experimental techniques and the relationship between the laboratory and industry. Again the tradition of the old Cavendish is as a laboratory with little interest in either advanced techniques or industrial collaboration, and again this tradition, captured in the 'string and sealing wax' image, has a lot of truth in it. But just as the laboratory was becoming more sophisticated in its use of theory in the late 1920s and early 1930s so it was also becoming technically more advanced. We therefore take a look at the development of new instruments in this period and also at the growing links with industry, in the form of the Metropolitan-Vickers Company.

The recollections in Part 3 are opened by Nevill Mott, who was working in Fowler's group in 1932 and recalls the relationship between theory and experiment in the work of the laboratory at that time. Alan Wilson, then lecturing in the Mathematics Faculty, describes the institutional background to Cambridge theoretical physics in the 1920s and 1930s. W B Lewis, who played a major part in the development of new techniques and instruments, recalls aspects of this development. Vivien Bowden, who in 1932 worked with both Lewis and the other main innovator Wynn-Williams, draws attention to the chance coincidences without which the instruments of nuclear physics and thus the subject itself would not have been possible. Wynn-Williams himself discusses his invention of the scale-of-two counter. Thomas

Allibone recalls the very close connections between the Cavendish and Metropolitan-Vickers and the way in which these operated during the Cockcroft–Walton project.

As we have already noted, the results of 1932, great as they were, were not unique in the Cavendish of the 1930s. In Part 4 of this volume we therefore move forward to the events, both published and unpublished, of 1933–4, and in so doing we also give a picture of how Cambridge appeared to young European physicists visiting it for the first time. Following a short introduction in which we note some of the changes that took place in physics and the Cavendish in the mid-1930s, we include three sets of recollections. In an extract from his book on Rutherford, published some years ago, Mark Oliphant recalls their work together in 1933 which lead to the successful demonstration of deuterium fusion. Maurice Goldhaber, who left Germany as a student that same year, pays tribute to the work of Chadwick and recalls his first impressions of Cambridge and his work with Chadwick on the nuclear photoelectric effect. Rudolf Peierls, another of the refugee scientists, then recalls his impressions of Cambridge formed while working at the Cavendish in 1933 and 1935–7.

Notes

1   Rutherford to Bohr, 21 April 1932, *SHQP*

Part I

Three Famous Experiments of 1932

1.1 Introduction

Chadwick discovers the neutron

The suggestion that there might exist neutral particles, generally conceived of as condensed hydrogen atoms or bound states of the two established elementary particles, a proton and an electron, had been put forward many times before such particles were actually discovered in 1932. The idea of neutral particles can, indeed, be traced back to the last century.[1] The story of what we now know as the neutron is most conveniently begun, however, with the prediction of its existence by Rutherford in 1920.

It was of course Rutherford who had made nuclear physics possible. At the beginning of the century he had been one of the pioneers of the study of radioactivity. Then in 1911 he had demonstrated, by bombarding atoms with alpha-rays and observing the scattering of these rays onto zinc sulphide scintillation screens, that the atom behaved as if its mass were concentrated in a small central positively charged core. With this famous and brilliant experiment the modern solar-system model of the atom, soon to be augmented by Bohr's quantum hypothesis of 1913, was born. The alpha-particles themselves were recognised as helium nuclei and the nuclear atom became the basis of atomic physics.

After the Great War, the nature of the nucleus was further clarified. In 1919 the Cavendish physicist F W Aston demonstrated, in a series of experiments with his newly designed mass spectrometer, the existence of isotopes, or atoms of the same chemical substance but with different nuclear masses, which had been predicted a few years earlier by Soddy.[2] Aston's measurements confirmed that the masses of nuclei were approximately multiples of the mass of the proton or hydrogen nucleus, and for the next fifteen years the nucleus was thought to be composed of protons and electrons, the number of protons determining the atomic weight and the difference between the number of protons and that of electrons determining the charge of the nucleus and thus the chemical nature of the atom. At the same time Aston also measured accurately the small difference between the mass of a compound atom and the sum of the masses of its components. Since an atom had a stable existence it had to have less energy than would be required for its parts to be set free, and given the established equivalence between energy and mass this meant that its mass had to be smaller than the sum of the masses of its parts. Aston demonstrated that this mass defect or binding energy was in fact quite significant, and this opened the way for the possibility of the energy-releasing fusion of nuclei.

Concurrent with Aston's experiments, Rutherford, in Manchester, pursued some results obtained by his colleague Marsden before the war.[3] Bombarding hydrogen, oxygen and nitrogen with alpha-rays he was able to measure the effective radii of their nuclei and to show that these were too small, and the components too distorted, for the nucleus to be describable in terms of the known gravitational and electromagnetic forces. The anomalous results he obtained with the bombardment of nitrogen were, in their way, more startling still. In one of the most impressive exercises ever recorded in experimental physics he finally discovered these anomalies to be due to protons ejected from the nitrogen atoms. The nuclei of nitrogen had fused with the incoming helium nuclei to form nuclei of an isotope of oxygen with the release of a proton. For the first time ever an artificial transmutation of an atom had been observed.

Following up these results in 1920, by which time he had moved to Cambridge as the new director of the Cavendish Laboratory, Rutherford also observed what he took to be helium nuclei of mass 3. This interpretation was in fact erroneous, as he discovered the next year, and he had in fact been observing ordinary helium 4; but the mistake led to a fruitful path of speculation which he offered tentatively to the world in his Bakerian Lecture to the Royal Society in 1920.[4] If there could exist stable helium nuclei of mass 3, interpreted as being composed of three protons and one electron, then there might also exist, he suggested, stable combinations of two protons and one electron (deuterium or heavy hydrogen nuclei), and of one proton and one electron. Shortly afterwards, he christened the latter neutrons.[5]

This speculation, tentative as it was, might have been abandoned when Rutherford discovered his original mistake; but the concept of the neutron had a very strong appeal. Considerable concern was being caused at the time by the two connected problems of how the stars acquired their energies—a process for which a subatomic energy source was now beginning to be seen as necessary—and how the heavier elements were generated, presumably within the stars, from the lighter elements. The first question had been taken up by the astronomer Eddington, Rutherford's colleague at Cambridge, who had himself drawn on Rutherford's and Aston's results to suggest the possibility of fusion, entailing a building up of elements, as a source of stellar energy.[6] The latter problem occupied the attention of Rutherford's former student Niels Bohr, and was also of particular concern to Rutherford himself. He no longer had any empirical evidence for the existence of the neutron, but he had convinced himself that its existence was necessary if the building up of the heavier elements was to be satisfactorily explained. A programme of searching for neutrons experimentally was accordingly undertaken at the Cavendish. James Chadwick, who had moved down from Manchester with Rutherford, was one of those involved.

Throughout the 1920s the experimental search for the neutron continued, and this is described below by both Feather and Chadwick. By 1931 theoreticians too were taking the possibility of the neutron seriously. But still it was not found. Meanwhile, however, as Feather records below, investigation of the effects of alpha-ray bombardment had continued unabated both in Cambridge and on the continent of Europe, and new electrical counting methods had been introduced to supplement the existing scintillation screens and cloud chambers. In 1930 Bothe and Becker observed some unexpected penetrating radiation which they assumed to be gamma-radiation, i.e. light-quanta, emitted when beryllium was bombarded by alpha-rays from a polonium source. Soon afterward, Webster, working under Chadwick's direction in Cambridge, independently observed a similar phenomenon. In June 1931 Chadwick and Webster actually considered the possibility that the penetrating radiation might be neutrons, but were unable to establish anything experimentally. Then in early 1932, by which time Webster had moved to Bristol, Irène and Frederic Joliot-Curie (the daughter and son-in-law of Marie Curie) reported in Paris that the beryllium radiation was even more penetrating than had been thought, and compared it with the recently investigated cosmic rays also thought—though not, as we shall see, by everyone—to be high-energy gamma rays.

In a second paper published a few weeks later the French team reported their attempts to determine whether the behaviour of the rays matched that reported for cosmic rays by the American Robert Millikan, who had recently visited their laboratory. They still assumed the penetrating radiation to be gamma-radiation, but the sums for the collision effect they reported just did not add up, and when the paper reached Cambridge Chadwick and Rutherford were immediately convinced that something was wrong. As Feather and Chadwick himself describe below it did not take long for Chadwick to show that the penetrating rays were in fact material ones and not light-quanta, and that the results pointed to their being the long-sought-for neutrons. A suitable polonium source of alpha-rays had recently been prepared (for which see also the recollections of Duncanson in Part 2 below). An electric counter sensitive to neutrons and a small ionisation chamber linked up to a valve amplifier and thus to an oscillograph, had also been developed.[7] It was a straightforward exercise for Chadwick to determine the range of protons ejected from hydrogen (in the form of paraffin wax) under the bombardment of the penetrating beryllium rays, and to measure the ionising power of recoil atoms produced when these rays were passed through a variety of different gases. The results were quite inconsistent with the hypothesis that the rays were light-quanta, but quite consistent with their being massive particles with neutral charge and mass very close to that of the proton.

On 26 January 1932 Chadwick had discussed at the Kapitza Club[8] the first of the Joliot-Curie's papers, noting the strangeness of their results but without drawing any startling conclusions. Just a fortnight later, on 10 February, he sent off a paper to *Nature* announcing the discovery of the neutron.[9] By 23 February, when Chadwick again addressed the Kapitza Club, Norman Feather had already started his investigations of the collisions of neutrons with nitrogen nuclei using a cloud chamber.[10] Philip Dee, as he himself records below, then joined in the project and studied the neutron–electron interaction. On 10 May all three physicists submitted papers to the Royal Society[11] describing their work in detail. There was at first a little concern that Dee's work did not show up the electron recoil tracks anticipated, but Harrie Massey at the Cavendish and Niels Bohr, who discussed the neutron at length during an international seminar series at the Institute of Theoretical Physics in Copenhagen in April, quickly confirmed that the observations did in fact accord with theory.[12] Chadwick and Rutherford had never had any doubts anyway.

Cockcroft and Walton split the atom

Throughout the twenties physicists continued to use alpha-particles emitted from naturally occurring radioactive substances as their main tools for exploring the nucleus, but there was also a growing recognition that new artificial sources of high energy particles would be needed for further progress. In particular, while the naturally occurring alpha-rays had proved very useful for the study of the light elements, the investigation of heavier elements had not really been possible. The heavier elements had greater numbers of orbital electrons screening their nuclei and, more crucially, greater positive charges on the nuclei themselves. The electromagnetic repulsion between the nuclei and incident alpha-particles created a barrier between them, and it seemed that the natural alpha-rays had insufficient energy to overcome this. Moreover the natural sources were weak in terms of the numbers of alpha-rays produced, and even if their energy were sufficient to overcome the nuclear barrier the chances of their going near enough the nucleus to do so were slight, so that even the investigation of the light elements was severely limited. To make significant progress it therefore seemed as if both high energy and intense artificial projectile beams would be needed.

Since the only effective supply of alpha-rays was from the weak natural sources already used, the development of alpha-ray sources did not seem promising. But there did exist artificial sources of electrons and protons which offered currents many thousands of times stronger than those obtainable from radioactive substances, and attention naturally focused around

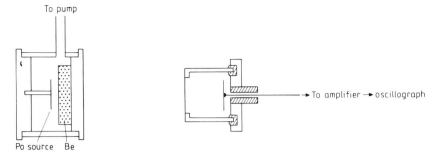

To pump

Po source   Be

To amplifier → oscillograph

Figure 1.1.1 Chadwick's source vessel and ionisation chamber (redrawn from *Proc. R. Soc.* A **136** 695 (1932)). When paraffin wax was placed just in front of the ionisation chamber, the range of emitted protons could be measured by inserting additional absorbing sheets of aluminium foil betwen the paraffin and the counter. The ionising powers of recoil atoms from other substances were measured by placing the test substances very close to the ionisation chamber, this time sealed with very thin gold foil, and observing the magnitudes of the deflections produced.

Figure 1.1.2 The Copenhagen seminar, 1932. On the front row, left to right are Bohr, Dirac, Heisenberg, Ehrenfest, Delbrûck and Meitner. Courtesy Professor A Hermann.

these. The basic problems were to generate at reasonable cost a high enough voltage to enable accelerated particles to penetrate the nuclear barrier, and to develop apparatus strong enough to withstand both these high voltages and the high vacua necessary to avoid the side effects that tended to result from them. Although the natural alpha-rays were weak in terms of current, they did produce rays with energies well in excess of those then obtainable artificially, and the problems were therefore daunting.

The first physicists to make significant inroads into these problems had been Merle Tuve and his supervisor Gregory Breit at the Department of Terrestrial Magnetism at the Carnegie Institution in Washington.[13] Tuve and Breit had met in Minnesota in 1923 while Tuve was still a student. Both had been interested in the production of high voltages for nuclear studies and they had discussed the matter between themselves and with another student, Ernest Lawrence. In early 1926 Tuve had actually applied for a fellowship that would have taken him to the Cavendish, with the intention of trying to build a high-voltage electron source under Rutherford's guidance; but he was persuaded to work with Breit instead. They therefore set out together to create high-energy proton and electron beams with which to probe further into the properties of the nucleus.

To obtain the necessary voltages easily and at reasonable cost and compactness, Tuve and Breit adopted a Tesla coil transformer capable of generating up to five million electron volts (5 MeV) for extremely short periods. The voltage was much higher than any known electron or proton source could take, but the established authority on such sources, Coolidge, was able to supply them with discharge tubes that could convert applied voltages into electron beams at energies up to 300 000 volts. By linking these in series they were able to build an apparatus that could accelerate electrons to just under 1 MeV. This compared unfavourably with the typical 3 MeV of electrons emitted in natural radioactive decay, but although the current, limited by the transient operation of the Tesla coil, was relatively weak, it was still equivalent to that produced by about 150 kilograms of radium, an enormous intensity by existing standards.

Concurrent with the work of Tuve and Breit, a similar attack on the problem of high-energy particle sources, again using a Tesla coil, was also made at the Cavendish by T E Allibone, who was seconded at his own request from the Metropolitan-Vickers electrical company to study for a doctorate under Rutherford in 1926.[14] In his Presidential Address to the Royal Society in 1927 Rutherford used this work, and specifically the achievements of Breit and Tuve, to draw attention to the need for artificial beam sources and to indicate the state of play.[15] If Rutherford's address was in part a statement of what had been achieved, however, it was also a statement of the limitations of this achievement. By 1930 Tuve had achieved both electron and proton beams at energies up to 1.2 MeV, but the Tesla coil was too spasmodic, unreliable and weak an energy source for these beams to be accurately controllable and put to use in nuclear research.[16]

Tuve did eventually produce a proton beam with which to disintegrate the light elements, but this was only after meeting Robert Van de Graaff in 1931 and substituting the latter's electrostatic generator for the Tesla coil.[17] But he was not the first to achieve this, and his apparatus, limited by the small current of the generator, was not to play an important part in future developments. At the California Institute of Technology, (CalTech), Charles Lauritsen used a cascade of traditional transformers linked in series to get a much stronger current of about 1 MeV by 1928,[18] and in 1933 he too achieved the disintegration results, but again he was not the first, and his work made relatively little impact on long-term developments.[19] More significant, however, was the work of Ernest Lawrence.

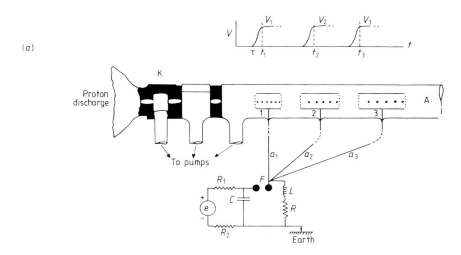

Figure 1.1.3 Linear accelerators for protons. (*a*) Ising's original design in which the electrons are periodically raised to a high voltage. (*b*) Wideröe's design, in which a rapidly oscillating current is passed across alternate electrodes, accelerating the protons between them.

Although he had long been interested in the problems of artificially accelerated beams, Lawrence only started to work actively in the field in 1928, when he stumbled across some work by Wideröe on the linear acceleration of particles. The basic idea behind Wideröe's work, which had in fact been expounded by Ising in 1924, was to take a stream of protons and pass them through a series of cylindrical metal electrodes in an evacuated chamber.[20] If carefully timed high voltages could be applied to the electrons using a current oscillating at an appropriate frequency, the electromagnetic field created would serve to accelerate the protons through them, gathering speed at each stage. Such an arrangement had the great advantage that it overcame the problem of having to generate very high voltages beforehand. The energy of the beam could be built up instead by the successive application of a relatively modest voltage. Such a principle had in effect already been used by Tuve and others, but their equipment had been limited by the number of discharge tubes that could be connected in series and by other severe technical limitations, whereas in the Ising–Wideröe set-up it seemed possible in principle to add any number of electrodes.

However, Lawrence found that a possible problem related to this apparatus was its length. On seeing Wideröe's work Lawrence had immediately calculated how many electrodes would be needed to accelerate protons to the order of 1 MeV, and what apparatus would be required to contain them. The result was an impracticably long tube. But he then asked, in a stroke of elementary genius, whether it might be possible to bend the proton beam around using a magnetic field, and send it in a circle through the same two or three electrodes repeatedly. The problem of timing the apparatus, so that the right field was present at the right instant within each electrode, had to be solved, but it worked out all right and within a day he had invented the cyclotron, or 'whirling device'.[21]

By 1930 Lawrence and his assistants at the University of California at Berkeley had constructed a cyclotron capable of accelerating protons in strong beams up to 700 000 volts, and when Cockcroft and Walton announced their disintegration results in 1932 the Berkeley team found that they already had the capacity to do likewise.[22] The credit for the splitting of the atom went, however, to Cockcroft and Walton, and the way in which they beat the considerable opposition we have described is most interesting. The Cockcroft–Walton project began when the young Walton arrived as a research student at the Cavendish in 1927 eager to work on a particle accelerator, and found Rutherford equally eager to have someone working on one. Walton himself recalls below the early phases of the resulting research programme, in which he tried first a circular device related to the modern betatron and then a linear accelerator, but both without success.

The change to a programme that would ultimately be successful followed a visit to the Cavendish by the young Russian physicist Georg Gamow. While the development of experimental physics had proceeded apace during the 1920s that of theoretical physics had been even more rapid, and had resulted in the creation in the middle of the decade of the science of quantum mechanics. It had already been known for some time that light possessed the paradoxical property known as wave–particle duality. In some circumstances it behaved as a diffuse wave form and in others as localised particles or light-quanta. In 1923 Louis de Broglie had presented a speculative theory in which matter too shared the wave–particle duality, and by the time quantum mechanics was created a number of experimental results were already being interpreted in terms of a wave property of electrons.[23] In particular some experiments by Davisson and Kunsman were being interpreted in terms of electron diffraction, while others by Ramsauer seemed to entail some kind of a barrier penetration phenomenon.[24] Electrons with insufficient energy to pass through an atom on the classical particle theory in fact had a finite probability of doing this. These wave properties of matter were in fact predicted by the new quantum mechanics, and in 1928 Gamow on one hand and Gurney and Condon on the other applied the new theory to radioactive processes.[25] On the classical theory a serious problem had been how alpha-particles could be emitted from an atom when they apparently had insufficient energy to surmount the potential barrier holding them to the nucleus. On quantum theory, however, they had a finite chance of breaking through the force barrier even if their energies were much smaller than would have been required by classical theory. Gamow worked out the probabilities for the radioactive emission of alpha-rays of different energies and showed that they were in fact in accord with the observed phenomena. Then in a second paper he reversed the situation and applied the same analysis to the penetration of a nucleus by alpha-rays, predicting theoretically the results obtained experimentally at the Cavendish for the transmutation of light elements as a result of alpha-particle bombardment.[26] In November 1928 Gamow sent a typescript of his second paper to Rutherford in anticipation of a personal visit in the New Year. The paper found its way to Cockcroft, who was then sharing a room with Walton and Allibone, and who quickly saw its implications. Even on the quantum theory, the energy required for alpha-particles to have a significant chance of penetrating the nucleus was considerable, but that for protons was much lower, and since the behaviour concerned was probabilistic in nature the low chance of a penetration at relatively low energies could be compensated for by high beam intensities. Having calculated the probabilities concerned Cockcroft proposed to Rutherford that the use of a proton accelerator operating at only 300 000 volts but with a fairly high current should pay dividends.[27] In January 1929 Gamow arrived in Cambridge and spoke on his theory at the Kapitza Club.[28] He confirmed Cockcroft's calculations on the probability of a proton beam such as that proposed successfully disintegrating a light element such as lithium, and with Rutherford's blessing a project was started with this aim in mind.

The Probability of artificial Disintegration by protons.

On Gamow's theory, the probability of an α particle entering the nucleus after coming within the effective collision radius r_m is

$$W_1 = e^{-\frac{16\pi e^2 z}{hv} J_K}$$

z being the atomic no. of the bombarded material, v the velocity of the α particle and J_k being a function of

$$k = \frac{m r_m v^2}{4\sqrt{3} e^2 z}$$

given on the appended curve. r_m is taken to be the radius at the peak of the potential energy curve of the α particle in the field of the nucleus, and is taken as $1.21 \ 10^{12} (\text{Atomi Wt.})^{\frac{1}{3}}$.

For a proton disintegration taking account of the half charge the probability becomes

$$W_1 = e^{-\frac{8\pi e^2 z}{hv} J_k'}$$

J_k being modified by an increase of $2\sqrt{2} kK$ to take account of the half charge.

The calculated probabilities are given below.

Volts.	α particle ± 1.	Proton ± 1.	Proton Boron.
3.10^6	0.20	1.00	1.00 ·
1.10^6	10^{-6}	0.062	0.55
5.10^5	10^{-15}	10^{-5}	0.055.
3.10^5	10^{-26}	10^{-8}	0.0059
2.10^5			$2.27.10^{-4}$

Thus a 300 k.v. proton should be 1/30th as efficient as a Polonium α in Boron

particle. The range of a 300k.v. proton is 5mm in air. Taking the no. in Al.

of disintegrations in Al. by Polonium as 10^{-6} of the incident α's we should expect $1.8 \ 10^6$ disintegrations per microampere of protons,

per mm of air equivalent.   JCC.

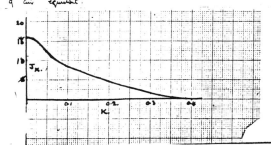

Figure 1.1.4 Cockcroft's memorandum to Rutherford, on the basis of which the Cockcroft–Walton accelerator was built. Reproduced from the Rutherford correspondence in *SHQP*, courtesy AIP Niels Bohr Library.

Figure 1.1.5 Cockcroft and Gamow working together in about 1930. Courtesy Cavendish Laboratory, University of Cambridge.

The story of the Cockcroft–Walton accelerator is told below by Walton and in Part 3 by Allibone, the latter from the viewpoint of Metropolitan-Vickers where a lot of the necessary development work was done. In 1930 Cockcroft and Walton actually got close to the intended voltage, reaching 280 000 volts, and used their beam to bombard beryllium and lead.[29] Gamow wrote impatiently in October that he would be 'very interested to hear when you'll begin to disintegrate atoms',[30] but although they did put in a lithium target in May 1931 Cockcroft and Walton cut short their experiments when a larger and more convenient room became available, and redesigned their apparatus to a higher energy so as to be confident of reliable operation at the energies actually required. In this Cockcroft the engineer triumphed over Cockcroft the scientist. Work on the new apparatus proceeded slowly, but by early 1932 it was beginning to operate satisfactorily. Bowden, who was working in the next room, has recalled the scene:[31]

> The apparatus was the very devil to maintain and to work. For months at a time Cockcroft and Walton spent their lives on ladders looking for leaks in that Plasticine and rubbing it with grease to make everything air tight. I have

always thought that this was a very peculiar way to prepare to become world famous! In the end everything was working at once and the machine produced a beam of very fast protons which emerged from the bottom of the tube. The experiment found out how far the protons would go in air, how they behaved in a magnetic field and what colour the glow was and so on. I very well remember what happened next. Rutherford came into the lab one day. First of all he hung a wet coat on a live terminal and gave himself an electric shock which didn't improve his temper. Then he sat down and lit his pipe—he

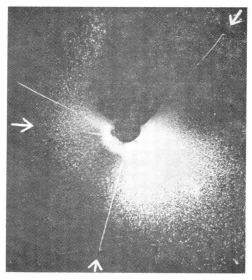

Figure 1.1.6. Dee's photographs of (*top*) the disintegration of lithium into two alpha-particles and (*bottom*) the disintegration of boron into three alpha-particles. Reproduced from *Proc. R. Soc.* **58** 628(facing) (1946).

always smoked very dry tobacco, so when he lit it, it went off like a volcano with a great big cloud of smoke, flames and piles of ash. Then he summoned Cockcroft and Walton and asked them what they were doing. He told them to stop messing about and wasting their time and get on and do what he'd told them to do months ago, and arrange that these protons were put to good use.

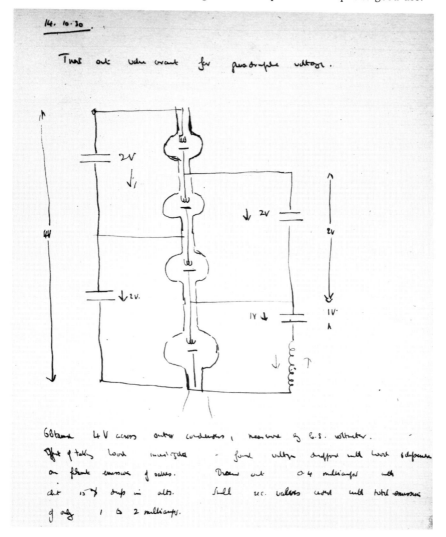

Figure 1.1.7 Extract from Cockcroft's notebook. Reproduced by kind permission of Lady Cockcroft and the Master, Fellows and Scholars of Churchill College in the University of Cambridge.

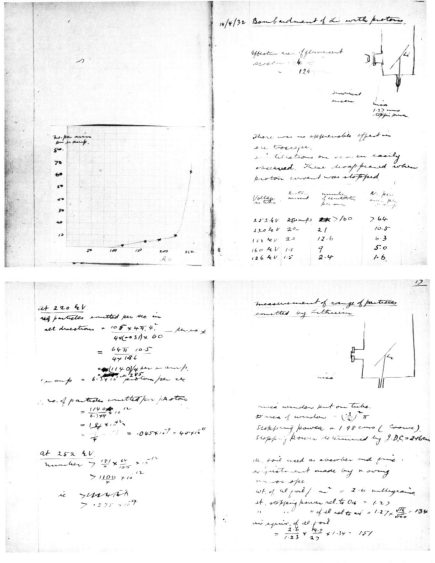

Figure 1.1.8 Extracts from Walton's notebook. Reproduced by kind permission of Professor Walton and the Master, Fellows and Scholars of Churchill College in the University of Cambridge.

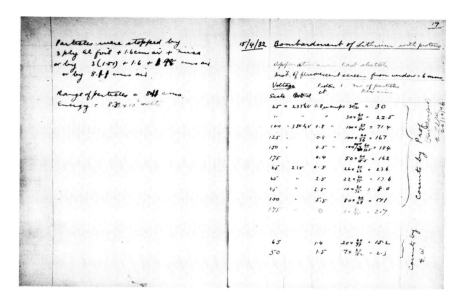

On 13 April 1932, according to Cockcroft's notebook, or 14 April according to Walton's, they finally observed, using a scintillation screen, the disintegration of lithium into two alpha-particles.[32] They were then quickly joined by Philip Dee, whose cloud chamber photographs confirmed beyond doubt that disintegration had been achieved, and the splitting of other light atoms soon followed.[33]

Blackett and Occhialini demonstrate the existence of the positron

Following the breaking of the electron–proton monopoly by Chadwick's discovery of the neutron, 1932 also saw the discovery of another new particle, the positive electron or positron. This discovery is generally attributed to an American, Carl Anderson, who interpreted photographs taken in the late summer of 1932 in terms of the new particle.[34] Anderson's results were not totally convincing, however, and credit for the conclusive demonstration of the positron's existence is usually given to Patrick Blackett and Giuseppe Occhialini, who carried out extensive investigations independent of Anderson at the Cavendish in 1932 and published their results early the following year.[35]

The experimental context of the positron work was not unconnected with that involved in the discovery of the neutron as described below by Feather, for the positron emerged from the study of the same cosmic ray phenomena as had led the Joliot-Curies astray in their interpretation of the radiation from beryllium. In the late 1920s the recently identified cosmic rays had been studied using cloud chamber observations by a number of distinguished physicists, including Robert Millikan in America, Lise Meitner in Germany and the Russian Skobeltzyn. In view of their high penetrating power the rays were generally assumed to be gamma-rays (light-quanta), but in 1929 Bothe and Kolhörster in Berlin put this assumption to the test.[36] As Feather recalls, Bothe was well in the forefront of the development of electrical counting methods for particle detection, and he was aided in this respect by his colleague Hans Geiger, whose work on counting devices had recently led to the invention of the Geiger–Müller counter, which was sensitive to charged particle radiations but not to gamma-rays. The use of a single counter could not provide a definitive answer as to the nature of the cosmic rays, as such a counter could easily be triggered by secondary electrons produced in the collisions between cosmic rays and terrestrial matter. But working on the assumption that any such secondary electrons would be significantly less penetrating than the original cosmic rays, Bothe and Kolhörster were able to provide *prima facie* evidence of the material nature of the latter by employing two counters with a layer of absorbing material between them. Doing this they observed coincidences between the responses of the two counters, which suggested that single material rays were passing through the entire system.

The German results were far from being definitive, but they did make a very strong impression upon the Italian physicist Bruno Rossi, then working in Florence.[37] With two students, one of whom was Giuseppe Occhialini, Rossi set about building some Geiger counters (then no easy matter) and repeating and improving the Bothe–Kolhörster experiments. In order to eliminate misleading near-coincidences he devised a coincidence circuit with which a mechanical counter could be activated only when a number of Geiger counters registered simultaneously. The coincidences were still observed, and Rossi came out in favour of a corpuscular hypothesis for the cosmic rays. In the autumn of 1931 an international conference on nuclear physics in Rome provided the opportunity for him to put forward his conclusions and debate them with the established authorities. The young Rossi's attack on the accepted orthodoxy cut little ice with Meitner, or indeed with Millikan who continued to uphold his own views in lectures at Paris and, on 23 November, to the Kapitza Club in Cambridge.[38] However the interest of the Cambridge physicists, several of whom, including Patrick Blackett, were present at the Rome conference, had been aroused.

While spending the summer of 1930 in Berlin, Rossi had struck up a friendship with Blackett, who was also visiting, and late in 1931 Rossi despatched Occhialini to Cambridge to learn from Blackett the techniques of cloud chamber work. Blackett was by this time the leading exponent of

Figure 1.1.9 Participants at the Rome conference, 1931. Photograph from the collection of Professor E Segré.

On the first and second steps, left to right, are Stern, Debye, Richardson, Millikan, A H Compton, Mme Curie, Marconi, Bohr, Aston, Bothe, Ellis, Rossi, Sommerfeld, Lise Meitner (hidden) and Goudsmit (at bottom of steps). Behind Sommerfeld is Rasetti, between Marconi and Bohr is Corbino and half hidden behind Marconi is Perrin. Behind Perrin are Beck and Fermi, with Ehrenfest looking ahead on the left. Between Stern and Debye is Brillouin, and behind him are the rest of the British contingent, Blackett on the extreme left being approached by Mott with hand in pocket, and in front of them Townsend.

cloud chambers in the world, and had himself looked briefly at cosmic rays. However, since expansions could only be made and photographs of the chamber taken at random, and with long delays between them, the study of cosmic rays had seemed to him unrewarding. Only a small percentage of the photographs he had taken had revealed any possible cosmic ray phenomena, and few of those were of any interest or use in attempting to explore the nature and significance of the rays. However, spurred on by the Rome conference, by Millikan's talk, and by Occhialini's arrival, Blackett set out to improve his cosmic ray results. Occhialini had come over to Britain for a few weeks to learn about cloud chambers, but he stayed for three years, and taught Blackett about Rossi's coincidence circuitry.[39] Blackett then devised a way to use a coincidence count between two Geiger counters, one above and one below a cloud chamber, to trigger an expansion with a photograph following at the optimum moment. Upon the first photographs from the new counter-controlled chamber being taken Blackett was found by Occhialini 'bursting out of the dark room with four dripping photographic plates held high and shouting for all the Cavendish to hear "one on each, Beppe, one on each" '.[40]

By the summer of 1932 Blackett and Occhialini could take photographs at the rate of one every two minutes, with an eighty per cent chance of a cosmic ray observation. Moreover, the counting mechanism was most likely to react when a cosmic ray collided in the chamber with a terrestrial atom, throwing off charged particles; and the more particles produced the more likely, again, that the mechanism would react and initiate an expansion. Blackett's observations were therefore biased in favour of the most interesting and informative cosmic ray phenomena. His apparatus did at this stage have one restriction, for whereas in the manually operated counter a large magnetic field could be turned on across the chamber to coincide with the expansion and give information as to the charges of the particles observed, the automatic chamber required a constant magnetic field, whose intensity was limited by the technology available. Nevertheless Blackett was able to set up a constant field at 3000 Gauss, and for the time being this was sufficient for his purposes.

The most interesting type of phenomenon, and one that had been discussed by Millikan in 1931, was the appearance of a pair of tracks apparently produced when a cosmic ray collided with something inside the cloud chamber. In one example discussed by Millikan two tracks curved almost symmetrically in opposite directions, apparently (but not unambiguously) from a common source. The implication was that either the two particles represented were oppositely charged, or they were travelling in opposite directions, with the coincidence of the paths being due purely to chance. Millikan interpreted his tracks as being due to an electron and proton emitted from a cosmic ray collision, and this was just about tenable, though the observed ionising power of the supposed proton suggested a much lower energy than did consideration of its curvature in the magnetic field. The alternative possibility that the track was in fact due to an electron moving in the opposite direction was also tenable, and such 'backwards electrons' had been noted by several experimenters during the previous few years; but while this agreed with the observed ionising power and curvature of the track the implied coincidence seemed unlikely.[41]

Carl Anderson was the first experimental physicist to suggest in public that neither of the explanations might be correct, and that there did in fact exist positively charged particles with the same mass as the electron. Anderson, who was a colleague of Millikan's, took a photograph in August 1932 in which a single particle was seen to cross a sheet of lead. The curvature and the length of the track combined to restrict it to something of mass comparable with an electron, but comparison of its two parts suggested three different explanations: that it must represent an electron leaving the lead with an energy three times that with which it went in, which made little sense, that the parts must represent different particles, which Anderson ruled out as too improbable in the circumstances, or that the track must refer to a positively charged electron, which Anderson accepted as the only reasonable explanation. Other subsequent photographs of the type discussed by Millikan, were then interpreted by Anderson using the positron hypothesis.[42] Meanwhile, however, Blackett and Occhialini had been working with the positron hypothesis very much in mind.

In developing the relativistic quantum theory of the electron, which showed that the electron spin was a necessary consequence of combining the quantum and relativity principles, Blackett's Cambridge colleague Paul Dirac had derived solutions to the field equations that seemed to correspond to electrons of negative energy. After he had unsuccessfully attempted to reconcile the solutions to these field equations with protons, and after Schrödinger, also without success, had attempted to eliminate them altogether, Dirac and fellow theoretical physicists such as Heisenberg and Oppenheimer had begun to think of the new solutions first as corresponding to 'holes' or 'vacancies' in a negative charge continuum, and then as positrons or 'anti-electrons'.[43] Though Dirac was not a member of the Cavendish his researches were well known there; he was a regular contributor to and member of the Kapitza Club and, as he recalls below, quite intimate at this time with Blackett. Blackett was therefore prepared from the beginning of his investigations for the possible existence of positrons, and his apparatus enabled him to test for that possibility far more rigorously than could Anderson or others. By August 1931 Blackett and Occhialini already had a hundred photographs,[44] and they quickly accumulated hundreds more, many including either pairs or showers of tracks. The latter were particularly useful, as the chances of backward travelling particles occurring could be virtually eliminated. The large numbers of examples allowed such possibilities in general to be properly treated. By the beginning of 1933 the cautious

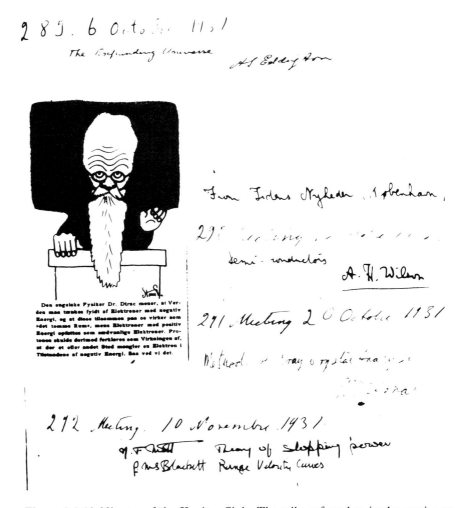

Figure 1.1.10 Minutes of the Kapitza Club. The talks referred to in the entries on these pages are:

A S Eddington The expanding universe

A H Wilson Semi-conductors

J D Bernal Methods of x-ray crystal analysis

N F Mott Theory of stripping power

P M S Blackett Range velocity curves

M L Oliphant Work of [?] on conductivity of thin metallic films
Robert A Millikan Cosmic ray possibilities
F C Powell Magnetostriction
E C Bullard Gravity observation
W L Webster Magnetism as a structure sensitive property
Reproduced from *SHQP* courtesy AIP Niels Bohr Library.

Blackett had finally convinced himself on the evidence he describes below that positrons were observed, and had in addition calculated that the observed frequency of the appearance of electron–positron pairs seemed to be consistent with that predicted by Dirac's theory.[45] Later analysis confirmed that this was so.

In April 1933 Blackett recieved a letter from Wolfgang Pauli suggesting that if both positive and negative electrons existed, and if the neutron existed, then perhaps it was not so fantastic to suppose the existence of the neutrino. This was a neutral particle with mass comparable to that of the electron, that Pauli had for the past few years been proposing unsuccessfully as a solution to the existing problems of nuclear structure.[46] He even suggested in a postscript that Blackett might have observed the effects of neutrinos. This he does not seem to have done; but later in the year Fermi's theory of beta-decay, partly anticipated by Pauli, opened the way for the acceptance of Pauli's neutrino.[47] A proton–neutron model of the nucleus had meanwhile been proposed by Majorana and Heisenberg, and the ingredients were then available for the modern theories of nuclear behaviour.[48]

Notes

1   Feather N 1960 *Contemp. Phys.* **1** 191, 257
2   Aston F W 1919 *Phil. Mag.* **38** 707; Aston F W 1920 *Phil. Mag.* **39** 449, 611; Soddy F 1910 *Ann. Rep. Chem. Soc.* 285; Soddy F 1913 *Nature* **92** 399
3   Rutherford E 1919 *Phil. Mag.* **37** 537, 562, 571, 581
4   Rutherford E 1920 *Proc. R. Soc.* A **97** 374; Rutherford E 1921 *Phil. Mag.* **41** 570
5   Glasson J L 1921 *Phil. Mag.* **42** 597
6   Eddington A S 1929 *Rep. Br. Assoc. Adv. Sci.* 34
7   See Part 3 below
8   Minutes of the Kapitza Club, Cockcroft Collection, Churchill College, Cambridge, and *SHQP* (see bibliography)
9   Chadwick J 1932 *Nature* **129** 312
10   Minutes of the Kapitza Club
11   Chadwick J 1932 *Proc. R. Soc.* A **136** 692; Feather N 1932 *Proc. R. Soc.* A **136** 709; Dee P I 1932 *Proc. R. Soc.* A **136** 727
12   Crowther J G 1974 *The Cavendish Laboratory 1874–1974* (London: Macmillan) p. 221; Bohr N 1961 *Proc. Phys. Soc.* **78** 1110
13   The account given here is based largely upon the essay by E McMillan in R Stuewer (ed) 1979 *Nuclear Physics in Retrospect. Proceedings of a Symposium on the 1930s* (Minneapolis: University of Minnesota Press) p. 111; see also Weiner C 1972 *Phys. Today* May issue p. 40

14 See Part 3 below

15 Rutherford E 1927 *Proc. R. Soc.* A **117** 300

16 Tuve M A, Dahl O and Breit G 1930 *Phys. Rev.* **35** 51; Tuve M A, Breit G and Halfstad L R 1930 *Phys. Rev.* **35** 66

17 Tuve M A, Halfstad L R and Dahl O 1933 *Phys. Rev.* **43** 942; Tuve M A 1933 *J. Franklin Inst.* **216** 1

18 Lauritsen C C and Bennett R D 1928 *Phys. Rev.* **32** 850

19 Crane R and Lauritsen C C 1933 *Phys. Rev.* **44** 783

20 Ising G 1924 *Arch. Mat. Astron. Fys.* **18** 1; Wideröe R 1928 *Arch. Elektrotech* **21** 387

21 Lawrence E O and Edlefson N E 1930 *Science* **72** 376; Lawrence E O and Livingston M S 1932 *Phys. Rev.* **40** 19

22 Lawrence E O Livingston M S and White M G 1932 *Phys. Rev.* **42** 150

23 See for example Gerber J 1969 *Arch. Hist. Exact Sci.* **5** 349; Born M 1978 *My Life* (London: Taylor and Francis); Jammer M 1966 *The Conceptual Development of Quantum Mechanics* (New York: McGraw Hill)

24 Davisson C and Kunsman C H 1923 *Phys. Rev.* **21** 637; Davisson C and Kunsman C H 1923 *Phys. Rev.* **22** 242; Ramsauer C 1922 *Ann. Phys., Lpz.* **64** 513; Ramsauer C 1922 *Ann. Phys., Lpz.* **66** 546; Ramsauer C 1923 *Ann. Phys., Lpz.* **72** 345

25 Gamow G 1928 *Z. Phys.* **51** 204; Gurney R W and Condon E U 1928 *Nature* **122** 439

26 Gamow G 1929 *Z. Phys.* **52** 510

27 Cockcroft J D Typescript memorandum *SHQP*

28 Minutes of the Kapitza Club

29 Cockcroft J D and Walton E T S 1930 *Proc. R. Soc.* A **129** 477

30 Gamow to Cockcroft, 21 October 1930, Cockcroft Collection, Churchill College, Cambridge

31 Lord Bowden in a lecture delivered at Canterbury University, New Zealand, 1979

32 Notebook in Cockcroft Collection, Churchill College, Cambridge

33 Crowther J G 1974 *The Cavendish Laboratory 1874–1974* (London: Macmillan) p. 222; Cockcroft J D and Walton E T S 1932 *Proc. R. Soc.* A **136** 619; Cockcroft J D and Walton E T S 1932 *Proc. R. Soc.* A **137** 229

34 Anderson C 1932 *Phys. Rev.* **41** 405

35 Blackett P M S and Occhialini G 1933 *Proc. R. Soc.* A **139** 699

36 Bothe W and Kolhörster W 1929 *Z. Phys.* **56** 751. For the historical background see Millikan R A 1930 *Nature* **126** 14, 29

37 Rossi B 1981 *Phys. Today* October issue 35

38 Minutes of the Kapitza Club

39 Blackett P M S 1969 *Rev. Nuovo Cimento* **1** xxxii

40 Occhialini G 1975 *Notes and Records of the Royal Society* **29** 144

41   See Hanson N R 1963 *The Concept of the Positron* (Cambridge: Cambridge University Press)
42   Anderson C 1932 *Phys. Rev.* **41** 405
43   Hanson N R 1963 *The Concept of the Positron* (Cambridge: Cambridge University Press)
44   Blackett P M S and Occhialini G 1932 *Nature* **130** 363
45   Blackett P M S and Occhialini G 1933 *Proc. R. Soc.* A **139** 699
46   Pauli to Blackett 19 April 1933, Blackett Papers, Royal Society: B134; see Brown L M 1978 *Phys. Today* September issue p. 23
47   Fermi E 1934 *Z. Phys.* **88** 161
48   Majorana E 1933 *Z. Phys.* **82** 137; Heisenberg W 1932 *Z. Phys.* **77** 1; Heisenberg W 1932 *Z. Phys.* **78** 156; Heisenberg W 1932 *Z. Phys.* **80** 587
49   See Stuewer R (ed) 1979 *Nuclear Physics in Retrospect* (Minneapolis: University of Minnesota Press)

1.2 The Experimental Discovery of the Neutron†

Norman Feather

In 1930 the nucleus was not yet 20 years old. Only 11 years previously Rutherford had moved from Manchester to Cambridge, his paper *An anomalous effect in nitrogen* just published.[1] He had written then, quite tentatively, 'If this be the case we must conclude that the nitrogen atom is disintegrated under the intense force developed in a close collision with a swift alpha-particle, and that the hydrogen atom which is liberated formed a constituent part of the nitrogen nucleus.' We now know that his flair for the correct assessment of experiment had not betrayed him: he had indeed observed, for the first time, the results of a man-contrived nuclear transmutation, but he went forward to Cambridge still not wholly convinced of his own success. 'I am hopeful that I will be able to settle the question definitely before long', he wrote[2] in January 1920; then, perhaps a little less confidently, in the following August 'I wish I had a live chemist tied up to this work who could guarantee on his life that substances were free from hydrogen'.[3] But he had been right all the time. The paper of Rutherford and Chadwick, of November 1921,[4] settled all reasonable doubts: from that moment the way to the experimental investigation of the inner constitution and structure of the atomic nucleus had been opened, straight and narrow though it was to remain for many years thereafter.

On any view, the problem of nuclear structure presented a challenge and an opportunity of immense magnitude, in 1921: in retrospect it is a matter of great surprise that no other physical laboratory in Britain, and none in North America, should have followed Rutherford in Cambridge, in an experimental attack on this problem in the following decade. Only in Vienna, in Berlin and in Paris was there any attempt to join in this adventure into the unknown. True, radium was still a rare commodity—and a strong source of radium was a first requirement for participation—but otherwise there was no elaborate experimental technique to be learned. Almost of necessity, if

† Extract from a paper read at the Tenth International Congress of the History of Science 1962, and first published in volume 1 of the Proceedings (Paris: Hermann et Cie, 1964) pp. 135–44.

it was desired to investigate a charged-particle radiation, alpha-particles or protons, of small intensity, in 1921, the investigator had to use the scintillation method of observation.[5] Even in 1930, in an experiment of fundamental importance in relation to the then newly-formulated wave-mechanical theory of identical-particle scattering, Chadwick[6] still used this primitive method as a matter of preference. In 1908 Rutherford and Geiger[7] had succeeded in recording the electrical effects of single alpha-particles, and in operating an electrical counting system consistently throughout an important research, but this was in the nature of a *tour de force*. Once they had satisfied themselves that the much simpler scintillation method gave the same results, they abandoned the electrical method: it was hardly used again in Rutherford's laboratory for 20 years.

In pioneering investigations, any technique ultimately reaches the limit of its usefulness. It is surprising, perhaps, that the scintillation method held the field as long as it did. Success depended upon the rigid self-discipline of the human observer. He became unreliable if he were used for more than two or three hours per day, or if within those times his periods of observation were not interrupted, every ten minutes or so, with longer periods of rest. He was generally not to be trusted if the counting rate were greater than 150 per minute or less than about 3 per minute. Yet this tedious and time-consuming method of observation remained Rutherford's invariable choice. In the field of x-ray studies others had already begun to look again to Geiger's electrical counter,[8] and the application of radiofrequency techniques to pulse amplification was a natural corollary. In the end, for Rutherford, it was failure to reach agreement as to matters of fact with the workers in Vienna,[9] a failure which in 1926 became almost complete,[10] that convinced him that the time had come to develop another method of single-particle detection, one which did not depend essentially on the human observer. Over the years 1927–30 this development, largely the work of Wynn-Williams,[11] produced the electrical recording devices for the next phase of the Cambridge attack.

One article of the Vienna heresy, as Rutherford saw it, was the statement that the light elements are copiously disintegrated by the alpha-particles of polonium.[12] These alpha-particles, of range 3.8 cm in air, are much less energetic than the alpha-particles of radium active deposit (of range 7.0 cm) with which Rutherford normally worked, or the alpha-particles of thorium active deposit (of range 8.6 cm) which he had used on occasion. The fact that in 1926 Chadwick observed a very small number of scintillations from aluminium when the radium active deposit alpha-particles were reduced in range to 3 cm by previous absorption in no way lessened the gravity of the disagreement. By and large, and within the normal sensitivity of the scintillation method, the light elements were not disintegrated by polonium alpha-particles, at any rate in Cambridge. Four more years had to pass before a polonium source was used in the Cavendish Laboratory, in such disintegration experiments, and by that time others, using electrical counters, had shown the way.

The electrical method of recording has three main advantages over the scintillation method, and a fourth to which we shall come in due course: it can be made continuous, it can accept particles at a much higher rate, and it also remains trustworthy at a much lower rate of counting than the 3 per minute of the scintillation method. To set against these advantages, the electrical counters of the early period had one serious disadvantage: they were much more sensitive to background gamma-radiation than were the zinc sulphide screens in which the scintillations were observed. Now, for equal alpha-particle intensities, the gamma-ray intensities of radium active deposit and polonium are as about $10^5:1$. Polonium sources, therefore, were very naturally at a premium when experiments with the new equipment were planned, and the stability of the equipment, even at very low rates of counting, soon made possible a subtantial harvest of results. Bothe and Fränz,[13] in Berlin, made the first notable contribution. Choosing boron, as the lightest element which was certainly known to be disintegrated by alpha-particle bombardment, they were able to show, in 1928, that there was abundant detail to be studied in the spectrum of the protons emitted from this substance when the alpha-particles of polonium were used for the bombardment. It took the whole of two years for the workers in Cambridge to make good the armoury of their equipment, and to rejoin the attack on equal terms. By 1930, when they had done so,[14] when electrical counting methods had been tamed, and polonium had been accepted as the alpha-particle source of the moment, the stage had been set for the experimental discovery of the neutron in 1932.

Reference has been made to the gamma-radiation. This penetrating radiation, of the same nature as the x-rays, had been slowly coming into its own, as an object of study, during the 1920s. Not until the beginning of the decade had it finally been established that this was a nuclear radiation. In 1922, Ellis[15] had been the first to show that well-defined states of nuclear excitation were involved in its emission. Then, in 1929, Rosenblum[16] had discovered the energy groups which make up the 'fine structure' of alpha-particle spectra, and Gamow[17], in 1930, had pointed to the obvious explanation. Alpha-particles may be emitted, possibly, with several different energies from identical nuclei, but when an alpha-particle is emitted with less than the maximum energy the product nucleus must be left excited, and in all probability a gamma-ray quantum will be emitted subsequently to restore the balance. Over the same period, 1929–30, it was becoming increasingly evident that the protons ejected from the light nuclei under alpha-particle bombardment were not all of the same energy, even when conditions were made as precise as possible:[18] it is not surprising, therefore, that experiments should have been mounted, in Cambridge and in Berlin, to look for gamma-rays in this situation, also. Again, for such experiments, it was obvious that only polonium alpha-particle sources would do.

The Berlin group were the first to publish. In August 1930 Bothe and Becker[19] gave a brief account of a survey of the penetrating radiations detected in a 'point' counter when the light elements from lithium to oxygen, inclusive, and magnesium and aluminium—and silver—were bombarded by the alpha-particles of polonium. No significant effect was found with carbon, oxygen or silver, but with boron, magnesium and aluminium the results were positive, as the authors had vaguely expected (for these elements are disintegrated with the emission of protons), only, more surprisingly, the results were also positive with lithium and beryllium. Indeed, with beryllium, the intensity of the penetrating radiation was nearly ten times as great as with any other element investigated. Furthermore, rough comparative measurements with a lead absorber showed that this beryllium radiation was notably more penetrating than the penetrating gamma-rays of radium.

Bothe and Becker published a more extended account of their findings in the following December.[20] They were commendably cautious in their conclusions. Admittedly, they had discovered a new and unexpected effect with beryllium, but they did not doubt that what they had observed was a gamma-radiation. Also, being unable to decide whether the quantum energy of this hypothetical radiation was greater or less than the maximum energy of the alpha-particles which produced it (realising that their absorption experiments were difficult of interpretation) they tended to accept the latter assumption, so minimising the novelty of the effect and making it possible to offer a tentative explanation with the least possible outrage to accepted ideas. Not until September 1931 did Bothe draw public attention to the fact, which was implicit in the earlier account, that when the hypothetical gamma-radiation from beryllium is produced (as it is) by alpha-particles of considerably less than the maximum energy of emission from polonium, then in that case its quantum energy is certainly greater than the corresponding alpha-particle energy. Only then[21] did he draw the inevitable conclusion that the alpha-particle must be captured whole by the beryllium nucleus if he were correct in his general assumption that the penetrating radiation is gamma-radiation.

In Paris, in 1930, not much attention had been paid to the new electronics. Up to that date there had been scant need of electrical counting methods in the Institut du Radium. The Curie laboratory possessed the largest supply of available polonium in the world, and the workers there had more experience than anyone else in its purification and use. It was little more than routine for them to prepare a polonium source almost ten times stronger than was possible elsewhere. With such a source Irène Curie was able to detect and study the beryllium radiation, using a simple ionisation chamber containing air at atmospheric pressure, and a sensitive electrometer. She gave a first account of her researches at the weekly meeting of the Academy of Sciences on 28 December 1931. Three weeks later she reported again in a joint paper with her husband. The reports were published promptly in the *Comptes Rendus.*[22, 23]

Curie's first publication reached Cambridge early in January 1932. She had carried the absorption experiment to greater thicknesses of lead than Bothe had done, and she had sharpened his comparison by using the more energetic gamma-rays of thorium to standardise her arrangement. She found the beryllium radiation to be even more penetrating than he had reported. Assuming, as he had done, that the radiation was gamma-radiation, she concluded that the quantum energy was between 15 and 20 MeV (at least three times the energy of the polonium alpha-particles), intermediate, as she said, between the energies of the most penetrating gamma-rays of radioactive origin and the cosmic rays. It can be stated, without bias, that this conclusion was uncritical on two counts. In the first place, it was already becoming abundantly clear that absorption in lead provided a very uncertain means of estimating quantum energy for high energy gamma-radiation. Experiments had shown[24] that even for quanta of 2.6 MeV energy there was a substantial contribution to the absorption for which there was no secure theoretical explanation, and therefore no method of predicting the likely magnitude of the effect at higher energies. In the second place, Curie's remark about the cosmic rays was entirely gratuitous, for there was no consensus of opinion regarding the nature of the primary cosmic radiation. Millikan had just previously (20 November 1931) been lecturing in Paris, advocating the view that the primary cosmic rays consist mainly of quanta of very high energy, but there were many eminent physicists who did not accept this view.

In Cambridge, in the interval between the receipt of Curie's first publication and her second, Webster sent to the Royal Society[25] an account of the experiments on which he had been engaged, with various interruptions, for the previous two years, under Chadwick's direction. He had covered much the same ground as Bothe and Becker, using a counter and electrical recording system for his earlier, and a high-pressure ionisation chamber for his later, experiments. On the assumption that the beryllium radiation was a gamma-radiation, he had made a reasonable attempt to estimate its quantum energy by investigating its absorption in aluminium and iron as well as in lead. On this basis the quantum energy was no more than 7 MeV. But Webster found one very puzzling fact. He discovered that when a beam of alpha-particles falls on a beryllium target the radiation which is emitted in the direction of motion of the incident alpha-particles is significantly more penetrating than that which is emitted in the opposite direction. It was very difficult to imagine any process involving gamma-ray quanta which could give an effect of this nature. If, on the other hand, the radiation were a particle radiation its kinetic energy would necessarily be greater forwards than backwards, but then the particles would have to be uncharged particles to explain their great penetrating power. Webster did not altogether neglect the possibility that he was dealing with a neutron radiation. Indeed, he made a half-hearted attempt to examine this possibility by enlisting the support of a colleague who took a few dozen expansion chamber photographs which in the upshot were not preserved. Nothing unusual was found. Current theoretical thinking inclined to the view that a neutron might ionise very sparsely if, as was assumed, it possessed a magnetic moment of sufficient magnitude. Carlson and Oppenheimer[26] and Langer and Rosen[27] had recently published theoretical papers on the subject. Webster, therefore, was looking for very thin tracks in the expansion chamber, and he did not find them. He concluded 'it is justifiable, for the present at any rate, to assume the [beryllium radiation] to be electromagnetic in nature.'

Webster's paper had hardly been despatched to London when the second communication from Paris arrived in Cambridge. Curie and Joliot had been following further the line of thought which Millikan had initiated. Millikan had claimed that the high-energy cosmic-ray quanta produced disintegrations in the atmosphere. He had evidence, he said, of the production of protons and electrons in such processes. Curie and Joliot modified their arrangement in order to see whether protons were produced by the action of the beryllium radiation. They fitted their ionisation chamber with a thin window, and placed various materials close to the window in the path of the radiation. They found nothing, except with materials such as paraffin wax and cellophane which already contained hydrogen in chemical combination. When thin layers of these substances were close to the window, the current in the ionisation chamber was greater than usual. By a series of experimental tests, both simple and elegant, they produced convincing evidence that this excess ionisation was due to protons ejected from the hydrogenous material, and they estimated the maximum energy of these protons as about 4.5 MeV. Quite clearly, what they had observed was not the result of nuclear disintegration, but of a process of elastic collision between the entities constituting the beryllium radiation and the nuclei of the hydrogen atoms in the material traversed.

The investigation which Curie and Joliot had reported to the Academy of Sciences on 18 January 1932 was of a classical simplicity, and it had produced a result of great novelty and importance, but the account of it which appeared in the *Comptes Rendus* had a title which was far from justified by the facts. It read 'The emission of protons of high velocity from hydrogenous materials irradiated with very penetrating gamma-rays.' From the outset there was clearly no thought of any revision of view concerning the nature of the beryllium radiation which produced these effects, and in the paper itself there was no satisfying explanation of the effects themselves, only the remark that if it were really a case of an elastic collision between a gamma-ray quantum and a hydrogen nucleus (nuclear Compton effect) then the quantum energy mush be the order of 50 MeV, poor agreement indeed with the estimate of 15 to 20 MeV based on the absorption experiment.

To one who had been following these developments from outside Paris, the gamma-ray hypothesis of Curie and Joliot had by this stage lost all reasonable claim to serious attention. And Chadwick had not only been following the developments; he had been involved in them, as director of Webster's research. For him the implication was obvious: his suspicions had been well-founded when Webster had discovered the difference in penetration of the forwards- and backwards-emitted radiations from beryllium. The latest result of Curie and Joliot provided just another effect, and a more startling one, which would present no difficulty of interpretation if this penetrating radiation were to consist not of quanta, but of neutrons. The hypothetical neutron of Rutherford's 1920 Bakerian Lecture[28] was a condensed hydrogen atom: a neutron of atomic mass effectively one. This was the neutron for which, from time to time, Chadwick had been looking for more than eleven years. If the beryllium radiation were to consist of such neutrons, their kinetic energy would be some 4.5 MeV (the maximum energy of the protons of Curie and Joliot) and the difficulty of explaining an energy release of 20 MeV, or 50 MeV, when polonium alpha-particles of 5.3 MeV were captured by beryllium nuclei would disappear overnight. This time he must clinch the matter decisively.

The logistics of Chadwick's attack on the problem were entirely direct: he must establish the fact that the beryllium radiation produced elastic collision effects in all substances, not merely in hydrogenous materials: and he must compare the maximum energies given to nuclei of different mass in such collisions. It is a necessary consequence of the laws of conservation of energy and momentum that the way in which this maximum energy depends on the mass of the projected nucleus is different when the incident radiation is a neutron radiation (neutrons of mass 1) from what it is when the incident radiation consists of quanta (of about 50 MeV energy). Taking the maximum energy of the projected protons as the standard of comparison, the maximum energy of projected nitrogen nuclei, for example, would be almost exactly one quarter of this amount on the former supposition and roughly one thirteenth on the latter.

Such was the essence of the plan; its successful execution depended upon the material resources available. In this respect Chadwick was well poised for the attack. For two years he had been using the electrical counting method in order to determine the ranges, and so the energies, of the groups of protons obtained from aluminium and fluorine, and other light elements, under alpha-particle bombardment.[14,29] It was a relatively simple matter to examine the protons projected from a thin layer of paraffin wax by the beryllium radiation by the same method.

Chadwick's detector was a small ionisation chamber in which the positive ions were collected without gas amplification. All the amplification necessary to operate a sensitive recording oscillograph was provided by a valve amplifier. For the purpose in hand, such an arrangement was ideal. If the amplifier were carefully designed, it was possible to ensure that the magnitude of the oscillograph deflection was directly proportional to the amount of ionisation produced in the chamber. This is the fourth advantage of the electrical method to which reference has already been made. The energy of the recoil atom producing the ionisation could thus be calculated directly from the size of the deflection on the oscillograph record. By filling the ionisation chamber with various gases, helium, nitrogen, oxygen or argon, the maximum energies given to the nuclei of these atoms could be determined simply.

Chadwick, then, had the necessary equipment. Also, he had, at last, an adequate supply of polonium. I had spent the academic year 1929–30 in Baltimore, in the Department of Physics of the Johns Hopkins University. I had come to know Dr West, the Englishman in charge of the radium at the Kelly Hospital. In this private hospital they had some 5 g of radium in solution, and at that time they pumped off the emanation each day, roughly 700 mCi, and transferred it to a small glass 'seed'. Each day, as one new seed passed into circulation, one, on the average, passed back into store, its therapeutic value already severely diminished. Obviously these old seeds were too 'hot' to be thrown away, even in 1930, and there were several hundreds of them safely stowed away at that time. Together they contained almost as much polonium as was available to Curie and Joliot in Paris. With great generosity Dr Burnam, Director of the hospital, gave me the greater part of this material, in the knowledge that it would find its way in the matter of a few months to the Cavendish Laboratory, to which I was returning to work with Chadwick, as a junior colleague.

The electrical counting equipment and the polonium were both available and in good order. So within ten days Chadwick had measured the range of the protons under various conditions, had detected the recoil atoms of helium, lithium, beryllium, carbon, nitrogen, oxygen and argon, and had determined the maximum ionisation produced, when these recoil atoms were liberated in the gas in the ionisation chamber. It was obvious at once that the whole picture made sense numerically if the penetrating radiation from beryllium consisted of neutrons of mass roughly equal to the proton mass; it made nonsense numerically if it were supposed that a quantum radiation were responsible for the effects observed. On 17 February Chadwick wrote a first short account of his observations. This appeared, under the title 'Possible existence of a neutron', in *Nature* ten days later.[30]

Here the story could well end, for the neutron had indeed been discovered, but there is room for a postscript: time did not stand still in the month of February 1932 in Paris and Berlin. In Paris, Curie and Joliot reported again to the Academy of Sciences on the 22nd of the month. They had filled their ionisation chamber first with helium and then with air, adjusting the pressure so that the density was the same in the two cases. They had found that the ionisation current due to the beryllium radiation was very different in the two gases: nearly five times as great in the helium as in the air. This observation was the counterpart of Chadwick's direct observation of the ionisation due to individual recoil atoms of helium in his chamber. Curie and Joliot interpreted it in just this way. They had evidence of a different kind, though even less direct, for the projection of carbon nuclei by the radiation. They were led to believe, quite naturally, that the phenomenon of the projection of nuclei by elastic collision was a general one, as Chadwick had demonstrated by more numerous examples. The next number of *Comptes Rendus* carried their report in print.[31] It was entitled 'The absorption of very high frequency gamma-radiation by collisions with light nuclei'. Their point of view had not changed: for them the penetrating radiation from beryllium was still a high energy quantum radiation—and they postulated a new and unexplained mode of interaction of such radiation with atomic nuclei, a mode which was, so they concluded, the more intense the smaller was the nuclear charge of the nucleus concerned. They wrote it out in symbols, assigning the letter J to represent this new effect. At the least, it was an unfortunate choice of symbol for this hypothetical new phenomenon.

In Berlin, and then in Giessen, Becker and Bothe had been working steadily to good purpose. They had devised a simple and ingenious arrangement which was essentially a gamma-ray spectrometer specially designed for the study of gamma-rays of high quantum energy and small intensity. Two thin-walled counters, used in coincidence, provided the means of estimating the maximum energy of the Compton electrons produced by the gamma-rays, and so the quantum energy of the incident radiation. The method involved long periods of counting, but once the observations had been made there was no ambiguity in their interpretation. In this way they made more precise their earlier estimate, by the absorption method, of the quantum energy of the gamma-rays produced in the alpha-particle bombardment of boron—and more secure their earlier conclusion that these gamma-rays are emitted when the disintegration protons have less than the maximum energy. With beryllium, too, they found copious gamma-rays, and they estimated the quantum energy as 5.1 MeV. They showed that this quantum energy was essentially the same whether the gamma-ray quantum was emitted in the direction of alpha-particle incidence, or in the reverse direction, and the same whether the alpha-particles had 5.3 MeV energy or 3.5 MeV. By the time that they came to write their first preliminary report of their investigation, in April 1932, Chadwick's letter to *Nature* was already in their hands. As a result they were able to give a consistent account of the whole phenomenon for the first time;[32] the penetrating radiation from beryllium was, after all, a mixture of neutrons and gamma-rays, and there was every reason to suppose that the neutrons must consist of two groups differing in energy by the energy of the gamma-rays. Later research[33] confirmed this prediction: there was nothing essentially new in the scheme of things at all, except the basic novelty, that neutrons were emitted rather than protons, in this case, under alpha-particle bombardment. Thereafter the two processes were seen as intrinsically similar: neutrons and gamma-rays, or protons and gamma-rays, or both protons and neutrons and gamma-rays; this was to become the pattern of interpretation, in the future, of all disintegration experiments involving alpha-particles and light nuclei.

Our final comment can be brief. Rutherford's neutron had been found, yet on closer scrutiny it turned out not to be Rutherford's neutron precisely. It was not a condensed hydrogen atom. Though small in size, it was inflated in energy: its mass was greater than the mass of the neutral atom of hydrogen.[34] In the free state it was radioactive and of short lifetime.[35] In the sparse matter of a giant star it was hardly to be reckoned with as an agent of nuclear synthesis. Such are the hazards of speculation and the devious paths to discovery.

Notes

1 Rutherford E 1919 *Phil. Mag.* **37** 581
2 Letter to Stefan Meyer
3 Letter to Boltwood
4 Rutherford E and Chadwick J 1921 *Phil. Mag.* **42** 809
5 Regener E 1908 *Verh. Dtsch. Phys. Ges.* **10** 78
6 Chadwick J 1930 *Proc. R. Soc.* A **128** 114
7 Rutherford E and Geiger H 1908 *Proc. R. Soc.* A **81** 141
8 See for example Bothe W and Geiger H 1925 *Z. Phys.* **32** 639
9 See Kirsch G and Pettersson H 1926 *Atomzertrümmerung* (Leipzig: Akademische Verlagsgesellschaft)
10 Chadwick J 1926 *Phil. Mag.* **2** 1056
11 Wynn-Williams C E 1927 *Proc. Camb. Phil. Soc.* **23** 811; Wynn-Williams C E and Ward F A B 1931 *Proc. R. Soc.* A **131** 391; see also Greinacher H 1926 *Z. Phys.* **36** 364
12 The main disagreement between Vienna and Cambridge was much wider. The Vienna workers claimed to have observed disintegration with almost every element observed; in Cambridge only some of the lighter elements were observed to show the effect.
13 Bothe W and Fränz H 1928 *Z. Phys.* **49** 1
14 See Chadwick J, Constable J E R and Pollard E C 1931 *Proc. R. Soc.* A **130** 463
15 Ellis C D 1922 *Proc. R. Soc.* A **101** 1
16 Rosenblum 1929 *C. R. Acad. Sci., Paris* **188** 1401
17 Gamow G 1930 *Nature* **126** 397
18 Rutherford E and Chadwick J 1929 *Proc. Camb. Phil. Soc.* **25** 186
19 Bothe W and Becker H 1930 *Naturwiss.* **18** 705
20 Bothe W and Becker H 1930 *Z. Phys.* **66** 289

21   Bothe W 1931 *Phys. Z.* **32** 661
22   Curie I 1931 *C. R. Acad. Sci., Paris* **193** 1412
23   Curie I and Joliot F 1932 *C. R. Acad. Sci., Paris* **194** 273
24   Tarrant T G P 1930 *Proc. R. Soc.* A **128** 345; Chao C Y 1930 *Proc. Natl. Acad. Sci. USA* **16** 431; Tarrant T G P 1932 *Proc. R. Soc.* A **135** 223
25   Webster H C 1932 *Proc. R. Soc.* A **136** 428
26   Carlson J F and Oppenheimer J R 1931 *Phys. Rev.* **38** 1787
27   Langer R M and Rosen N 1931 *Phys. Rev.* **37** 1579
28   Rutherford E 1920 *Proc. R. Soc.* A **97** 374
29   Chadwick J and Constable J E R 1932 *Proc. R. Soc.* A **135** 48
30   Chadwick J 1932 *Nature* **129** 312
31   Curie I and Joliot F 1932 *C. R. Acad. Sci., Paris* **194** 708
32   Becker H and Bothe W 1932 *Naturwiss.* **20** 349
33   Maier-Leibnitz H 1936 *Z. Phys.* **101** 478
34   Chadwick J and Goldhaber M 1934 *Nature* **134** 237
35   Snell A H, Pleasanton F and McCord R V 1950 *Phys. Rev.* **78** 310; Robson J M 1950 *Phys. Rev.* **78** 311

1.3   Some Personal Notes on the Discovery of the Neutron†

James Chadwick

A few months after his Bakerian Lecture of June 1920, in which he first mentioned what had been in his mind for some time, the possible existence of a neutral particle formed by the close combination of a proton and an electron, Rutherford invited me to join him in following up the experiments on the artificial disintegration of nitrogen which he had made in Manchester.

There were a number of reasons for this invitation, so welcome to me. Among them was that I had made some improvements in the technique of counting scintillations, namely better optical arrangements and a strict discipline; but also he wanted someone to talk to, to while away the tedium of working in darkness.

It was during the periods of waiting to begin counting that he expounded to me at length his views on the problems of nuclear structure, and in particular on the difficulty in seeing how complex nuclei could possibly build up if the only elementary particles available were the proton and the electron, and the need therefore to invoke the aid of the neutron. He freely admitted that much of this was pure speculation, and, being averse to speculation without some basis of experiment, he seldom mentioned these matters except in private discussion. Indeed, I believe that only on one occasion after the Bakerian Lecture did he again refer publicly to his views on the role of the neutron. He had not abandoned the idea, and he had completely converted me. From time to time in the course of the following years, sometimes together, sometimes myself alone, we made experiments to find evidence of the neutron, both its formation and its emission from atomic nuclei. I shall mention some of the more respectable of these attempts; there were others which were so desperate, so far-fetched as to belong to the days of alchemy.

Immediately after the Bakerian Lecture Rutherford had asked J L Glasson to look for the production of neutrons when an electric discharge was passed

† Paper read at the Tenth International Congress of the History of Science, 1962, and first published in volume 1 of the Proceedings (Paris: Hermann et Cie, 1964) pp 159–62.

through hydrogen, and a little later J K Roberts made a somewhat similar experiment. He could not really have expected that any evidence of the neutron would turn up in this way, but it had to be tried. Both the mass of hydrogen and the voltages used in these experiments were quite trivial.

It seemed to me not too unreasonable to look at hydrogen in the normal state, notwithstanding its apparent stability. If a close combination of proton and electron were possible at all, it might take place spontaneously; and the neutron so formed might break up again under the action of the cosmic radiation. With Rutherford's approval, I tried, in 1923, to detect the emission of gamma-radiation from the formation of neutrons in a large mass of hydrogenous material, using an ionisation chamber and a point-counter as the means of detection.

A few years later, in 1928, Geiger and Müller devised what is now universally called the Geiger counter, which enormously increased the ability to detect gamma-radiation. Geiger very kindly sent me two of his new counters, as well as instructions for making them. Immediately, Rutherford and I used this new instrument to repeat the experiment with hydrogen. We went to all manner of tricks in the hope of finding some trace of the neutron. We also examined in the same way some of the rare gases, and any rare element we could lay our hands on, just in case any sign of the formation of the neutron or its emission might turn up. I mention these experiments only in a general way because some were quite wildly absurd.

After my first attempt in this way I had considered the possibility that the neutron could be formed, or exist, only in a strong electric field; and that perhaps one might find some evidence by firing fast protons into atoms, especially those of higher atomic number where some electrons were tightly bound. This was the vague idea behind the remark in a letter to Rutherford which is quoted in Eve's *Life* (p. 301)—'I think we shall have to make a real search for the neutron. I believe I have a scheme which may just work. . .'. I thought that at least 200 000 volts would be necessary for the acceleration of the protons. No suitable transformer was available and, although Rutherford was mildly interested, there was no money to spend on such a wild scheme. The research grant of the Cavendish was about £2000 a year, little even in those days for the amount of work which had to be supported. I persisted with the idea for a year or two, and in the intervals of other work I tried to find a way of applying Tesla voltages to the acceleration of ions in a discharge tube. I had quite inadequate facilities, and no experience in such matters. I wasted my time, but not the Laboratory money.

During our work on the disintegration of the lighter elements by alpha-particles Rutherford and I had not been unmindful of the possibility of the emission of neutrons, especially from those elements which did not emit protons. We looked for faint scintillatings due to a radiation undeflected by a magnetic field. The only specific reference to the search for the neutron in this way was made in a paper published in 1929, some years after the experiments.

The case of beryllium was interesting for two reasons. It did not emit protons under alpha-particle bombardment; and, though a false argument, the mineral beryl was known to contain an unusual amount of helium, suggesting that perhaps the Be nucleus split up under the action of the cosmic radiation into two alpha-particles and a neutron.

This matter intrigued me on and off for some years. I bombarded beryllium with alpha-particles, with beta-particles and with gamma-rays, generally using the scintillation method to detect any effect. In those days this was the only method of much use in the presence of the strong gamma-radiation of the radium active deposit, the chief source of radiation available to me. Quite early on, too early perhaps, I tried to devise suitable electrical methods of counting. I failed. Later, when the valve amplifier method had been developed by Greinacher, and put into use in the Cavendish by Wynn-Williams, I was also able to make a polonium source, small but just enough for the purpose. With Constable and Pollard, I had another look at beryllium, and for a short but exciting time we thought we had found some evidence of the neutron. But somehow the evidence faded away. I was still groping in the dark.

The first indication of the neutron came in the work of H C Webster on the gamma-radiation excited in beryllium by alpha-particle collisions. I had had such work, the excitation of gamma-rays by bombarding light elements with alpha-particles, in mind for some years. An attempt had been made by L H Bastings, but this failed, because the polonium source was too weak and the instrument of detection, the electroscope, too insensitive. When the Geiger counter became available Webster took up this quest, but his first efforts were not very rewarding—I was still short of polonium.

This deficiency was overcome by the kind intercession of Dr Feather, then in Baltimore, and the generosity of C F Burnam and F West of the Kelly Hospital. They sent me, first by the hand of Dr Feather and later by post, a number of old radon tubes which together contained what was, for me, a very large quantity of radium D and its product polonium. This gift was of immense value both immediately and later on.

In the meantime, Bothe and Becker had taken up this matter and they were the first to publish results. But Webster made a most interesting observation, that the radiation from beryllium which was emitted in the same direction as the incident alpha-particles was more penetrating than the radiation emitted in the backward direction. This fact, clearly established, excited me; it could only be readily explained if the radiation consisted of particles, and, from its penetrating power, of neutral particles. Believing that a neutral particle would produce tracks, though very sparsely ionised, I suggested that he should pass the radiation into an expansion chamber. To our dismay, for we were convinced that a neutral particle of some kind was involved, no such tracks were to be seen. We were very puzzled; we did not know how to reconcile the observations.

This near-missed occurred in June 1931. Shortly afterwards Webster left Cambridge for Bristol. I decided to take up the matter afresh, but my preparations were delayed, perhaps fortunately, by a change of my working quarters to another part of the laboratory. Then one morning I read the communication of Curie-Joliot in the *Comptes Rendus*, in which they reported a still more surprising property of the radiation from beryllium, a most startling property. Not many minutes afterwards Feather came to my room to tell me about this report, as astonished as I was. A little later that morning I told Rutherford. It was a custom of long standing that I should visit him about 11 AM to tell him any news of interest and to discuss the work in progress in the laboratory. As I told him about the Curie-Joliot observation and their views on it, I saw his growing amazement; and finally he burst out 'I don't believe it.' Such an impatient remark was utterly out of character, and in all my long association with him I recall no similar occasion. I mention it to emphasise the electrifying effect of the Curie-Joliot report. Of course, Rutherford agreed that one must believe the observations; the explanation was quite another matter.

It so happened that I was just ready to begin to experiment, for I had prepared a beautiful source of polonium from the Baltimore material. I started with an open mind, though naturally my thoughts were on the neutron. I was reasonably sure that the Curie-Joliot observations could not be ascribed to a kind of Compton effect, for I had looked for this more than once. I was convinced that there was something quite new as well as strange. A few days of strenuous work were sufficient to show that these strange effects were due to a neutral particle and to enable me to measure its mass: the neutron postulated by Rutherford in 1920 had at last revealed itself.

I trust that I shall not be misunderstood if I add a postscript to this story. It is unnecessary to record my satisfaction, and delight, that the long search for the neutron had, in the end, been successful. The decisive clue had indeed been supplied by others. This after all is not unusual; advances in knowledge are generally the result of the work of many minds and hands. But I could not help but feel that I ought to have arrived sooner. I could offer myself many excuses: lack of facilities, and so on. But beyond all excuses I had to admit, if only to myself, that I had failed to think deeply enough about the properties of the neutron, especially about those properties which would most clearly furnish evidence of its existence. It was a chastening thought. I consoled myself with the reflection that it is much more difficult to say the first word on any subject, however obvious it may later appear, than the last word—a commonplace reflection, and perhaps only an excuse.

1.4 Some Reminiscences of the Discovery of the Neutron

Philip Dee

A few of my personal reminiscences might be of interest, not only in respect of the events, but also of the attitudes and personalities of those who were closely concerned with the exciting early months of 1932.

The dominant question was, of course, the nature of the radiation emitted from beryllium under bombardment with alpha-particles. Curie and Joliot had observed the projection of protons, several MeV in energy, from paraffin subjected to this radiation and had interpreted the result as a Compton-type process of photon–proton collisions. There were, however, grave difficulties in accounting for the very high energies which such photons would have to possess and also the frequencies of such events.

Chadwick, using the same radiation from a Po/Be source, had made measurements, in proportional counters, of the pulses produced in various gases and had interpreted these pulses as being due to nuclei recoiling under the impact of neutrons. He found that the relative magnitudes of these pulses in different gases were compatible with a mass close to unity for an assumed neutron.

At a meeting of the Kapitza Club I heard Chadwick give an account of these experiments and it seemed to me that measurements of electrons recoiling from the beryllium radiation would very clearly distinguish between photons and neutrons because of the great difference in recoil energy to be anticipated on the two hypotheses. For photons of the assumed energy the electron recoil tracks would be several metres in length whereas for neutrons the electron recoil energy would be only a few keV—giving tracks perhaps 2–3 mm long.

At this time I was working at the Solar Physics Observatory with C T R Wilson, who worked there rather than at the Cavendish ostensibly to be free of radioactive background but really, I think, because he preferred to work independently and unharassed by other workers. We had been studying some basic condensation phenomena, but with the object also of perfecting a technique of operating a cloud chamber under extremely clean conditions.

We were at a stage where ion-counting by the droplet counting method was becoming quite accurate, the cloud chamber being almost totally free of any background condensation. I had recently been photographing photoelectron tracks from CuK x-rays—short electron tracks of just the type to be anticipated from the neutron collisions. I told Chadwick this at the end of his lecture and was delighted when he at once said that I could have the Po/Be source for overnight experiments, whilst he and Feather would be resting in preparation for their daytime labours.

It soon became clear, however, that no tracks of the desired type were to be found. Long tracks, such as would be expected on a photon hypothesis for the radiation, were sometimes present but it was soon shown that these arose from gamma-rays from the polonium source.

C T R had, I think, been a little unhappy at the interruption of our condensation work but had agreed, perhaps in tolerant understanding of a young man's desire to be involved in the exciting current topic. Soon, however, he had become so interested that he arrived one morning with all of his gold medals, including the Nobel Prize, for me to use to filter off the unwanted gamma-rays. These medals were so highly embossed that they stacked very badly, but I felt that I could not press him unduly when he demurred at my suggestion that they could easily be flattened to form a tidier pyramid. Occasionally some short electron tracks were observed but in every case these could be clearly identified as branches upon the few fast electron tracks still remaining.

During these few weeks I frequently met Rutherford and had to report my continued failure to find the desired electron tracks. At first his reactions were scathing. He took pains to emphasise that it should be 7.2 times as easy to find these tracks as to find the nuclear recoils. In self defence I used to show him photographs of nitrogen recoils which looked like trunks of trees and CuK photoelectron tracks which looked like fat tadpoles. After a while he accepted the situation and in later years would tease me about the niceties of cloud chamber operation. Rutherford was *absolutely certain* throughout that the neutron he had predicted so many years earlier had now been found. He once concluded a conversation with me by saying 'Well. . .they *have* to be there [the electron tracks]. . .we *have* to have it [the neutron]'.

I think that the *manner* of the discovery, involving the straightforward application of the laws of conservation of momentum and energy, so similar to much of his own earlier work, appealed to him above all other considerations, although he also felt strongly the need for a neutron as a universal nuclear constituent. Chadwick's reaction to my failure to find these tracks was quite different. He accepted at once that they did not occur, concluded that the neutron–electron collision cross section was very small and urged me to try to set an upper limit to its magnitude. Chadwick *knew* there was a neutron because of his *own* experiments.

I expect that many people would be surprised and perhaps amused that the negligible magnitude of the neutron/electron interaction was not generally recognised from the outset, but there was no direct evidence to this effect and even the theorists I consulted in the hope of obtaining some moral support were non-commital! However during March and April the neutron became accepted in the Cavendish as a definite, almost familiar, entity and the way was clear for the discovery and development of the tremendous range of neutron-based phenomena.

1.5   Personal Recollections of the Discovery of Fast Particles†

Ernest T S Walton

Tools have always had a fascination for me. As a boy and as a student, any money which came to me at Christmas and at birthdays was invariably spent on tools. This fascination arose undoubtedly from the power to do and make things which the possession of tools gave me. They could be used to put new ideas into concrete form and they could produce machines and instruments not available on the market.

Instruments and apparatus are the tools of the scientist. Every new instrument, if it is of any value, and every improvement in existing instruments and techniques opens a wider area to investigation. This may lead to important new discoveries such as the strange and totally unsuspected phenomenon of superconductivity which followed the development of low-temperature equipment. Holding such a viewpoint, it is not surprising that when given the opportunity of doing research, my mind has often turned to possible advances in technique. It is sometimes a matter of human interest to learn how people become involved in work which later attracts public attention. Hence more personal details may be excused on the ground that someone's curiosity may be satisfied.

While doing some research work soon after graduating, I found that the Royal Commission for the Exhibition of 1851 was offering valuable scholarships to enable young graduates to work at universities in the Commonwealth other than their own. Here was my chance. I was fortunate to be awarded a scholarship and fortunate also to be accepted by Rutherford as a research student. I say fortunate, because at that time Rutherford was at the height of his powers and was President of the Royal Society. He was Director of the Cavendish Laboratory, Cambridge—a laboratory made famous by a continuous line of distinguished professors of physics, Clerk Maxwell, Lord Rayleigh, Sir J J Thomson, and now Sir Ernest Rutherford. The laboratory

† This paper has been adapted by Professor Walton from his Nobel Lecture and from a talk broadcast in 1959, with the addition of more recently composed recollections.

had become the centre of attraction for physicists all over the world who were interested in radioactivity and atomic physics generally. The result was that in the laboratory were to be found many outstanding young physicists who later held a high proportion of the professorships of physics in the Commonwealth and not a few outside it. Furthermore, the laboratory was a place of pilgrimage for many famous men who would often give a lecture and have talks with research students working on lines in which they were interested.

I arrived at Cambridge in October 1927 and I had some difficulty in finding the famous laboratory for it was an unpretentious building tucked away inconspicuously up a side street and no passers-by seemed to have heard of it.

The usual practice in the Cavendish Laboratory was for new research students to spend the first term in an attic labelled by some 'The Nursery'. Here one learnt to make radioactive measurements and to do some glass-blowing and some high-vacuum work. During the first year the new arrival was encouraged to attend lectures. There was a rich choice available and I was thrilled to have the opportunity of attending lectures by people about whose fame and work I had read in the textbooks. First, there were Ruth-erford's own lectures in which he was always overflowing with an infectious enthusiasm for the subject. He was not always very well prepared for any mathematics which might arise but no doubt his very extensive duties of all sorts were responsible for any hesitations which became obvious. Equally great was the thrill of attending the lectures given by J J Thomson, famous for his research work on the passage of electricity through gases and for his discovery of the electron. He had been a most successful Cavendish Professor for 34 years and now he was Master of Trinity, Cambridge's largest college. C T R Wilson was Jacksonian Professor and was the inventor of the cloud chamber aptly described as the final court of appeal in matters of nuclear physics. F W Aston, whose name was almost synonomous with isotopes and the precise measurement of their masses, was still there. But I must refrain from mentioning any more of the eminent people there who contributed to the stimulating atmosphere in which the young physicist was placed. At the end of my term in 'The Nursery', the question of a line of research for me arose. Rutherford was, of course, fully alive to the importance of research suited to the individual and always tried to find something in which the student was really interested. He asked me about my own ideas for research. As an undergraduate, when reading about his own work on the disintegration of atoms by the use of natural alpha-particles, I wondered why other pro-jectiles were not also tried. By now I knew that alpha-particles were the only fast projectiles available to him. To produce other projectiles would have required the generation and application of electrical potentials of several millions of volts. This was much higher than anyone had previously been able to use in a laboratory. There was no point in suggesting to Rutherford that I should attempt to bridge such a large gap and so I put forward an indirect method not requiring the use of very high voltages.

I suggested to Rutherford the idea of accelerating electrons by letting them move many times round in a circular electric field as in the modern betatron. The suggested method of producing such a field was not very practicable. Rutherford pointed out that the time required was too long and stray fields would have time to cause particles to fly off the orbit. He suggested the use of a high frequency current in a circular coil. As he indicated, this would be much the same arrangement as used by J J Thomson in his work on the electrodeless discharge in gases. I used the only source of high frequency current then available to me, namely the trains of damped oscillations obtained from spark discharges. It was hoped that the alternating magnetic field combined possibly with a steady uniform magnetic field would constrain the electrons to move along the lines of the circular electric field. However, no trace of evidence was found for the presence of fast electrons.

Calculations showed that a uniform oscillating magnetic field was two times too strong to maintain electrons on the circular orbit and so they would spiral into the centre. The situation was even worse in the experimental arrangement used because the magnetic field increased on moving out in the radial direction. Detailed calculations showed radial stability could be produced by a magnetic field falling off inversely with radial distance if a suitable high frequency radial electric field was also present. Experiments along these lines failed because the arrangements used were too crude and no provision was made for stability in the axial direction. When I suggested this method to Rutherford I was unaware that just about two weeks previously in his presidential address to the Royal Society, he had stressed the importance of developing methods of producing particles of energies greater than alpha- and beta-particles. So perhaps he was glad to find someone anxious to work in this difficult new field! I was lucky that my suggestion was made at an opportune time.

In December 1928, I suggested to Rutherford another method for the production of fast particles. It was the method now called the linear accelerator in which a high frequency voltage is applied to alternate members of a line of cylinders. Rutherford did not appear to have heard of the principle but some rapid simple calculations which he made convinced him of its feasibility. Experiments on the method were not successful for two reasons. Only spark-produced high frequency currents were available and very little was known about the focusing of a beam of particles at a succession of gaps. Indeed, the ends of cylinders were covered with gauze in order to secure a field-free space inside the cylinders and this removed the focusing action now known to occur.

In 1929 it seemed that much development work would have to be done on the indirect methods of obtaining fast particles, while at the same time there were indications that, after all, nuclear disintegrations might be produced by particles of reasonably low energies. The application of the wave mechanics showed that there was a non-zero probability that particles might penetrate barriers which they could not surmount. If a sufficiently great number of low energy particles were directed against the barrier, some would go through. Using these ideas, Gamow was able to explain the Geiger–Nuttall law for the emission of alpha-particles. In 1928 he visited the Cavendish Laboratory, and Cockcroft discussed with him the reverse problem of getting particles inside a nucleus. As a result of this, Cockcroft sent a memorandum to Rutherford in which he showed that protons of only 100 000 eV energy had a small but not negligible chance of penetrating the nucleus of a light atom. As it was expected that very large numbers of protons could be accelerated by this voltage, an appreciable number of penetrations of the barrier should occur. This result had the effect of encouraging the transfer of attention from the indirect methods, which appeared to be a long-term project, to the production of fast particles by using high potentials of not unreasonable magnitude.

In the work carried out by J D Cockcroft and the writer, the aim was to accelerate particles by the direct application of potentials of up to 300 kV, these being about the highest potentials which it had been found possible to apply to a vacuum tube for the production of x-rays and cathode rays. The conventional tubes of the time were large glass bulbs with two stems and these were used both for the rectifiers and for the accelerating tube. The transformer voltage was rectified by the horizontal rectifiers placed in series, these being evacuated through a third bulb connecting them to the pumps. With this apparatus several microamperes of protons accelerated by about 280 kV could be obtained.

At this stage the laboratory used had to be vacated and a much larger room became available. Taking advantage of this, the production of much higher energy particles was attempted. For this purpose the voltage multiplier circuit shown in Figure 1.5.1 was used. It is a modification of one due to Schenkel. It gives a fourfold multiplication of voltage and is capable of extension to any even multiple of the transformer voltage. Essentially, it consists of condensers [capacitors] C_1 and C_2 in series, the voltages across them being maintained equal by means of the transfer condenser C_3. This condenser is connected, in effect, alternately in parallel in rapid succession across C_1 and C_2, C_1 becomes charged to twice the peak voltage (V) of the transformer because during one half-cycle, C_4 is charged to V through the rectifier D_1, while during the next half-cycle the voltage across C_4 is added to the transformer voltage and thus C_1 gets charged through the rectifier D_2 to twice the transformer voltage.

In addition to giving a steady voltage which may be any desired even multiple of the transformer voltage, the circuit has other advantages. It gives steady voltage tappings at intermediate points, these being useful when using a multisection accelerating tube. The rectifiers are all connected in series and so may be erected as a single column and evacuated by one diffusion

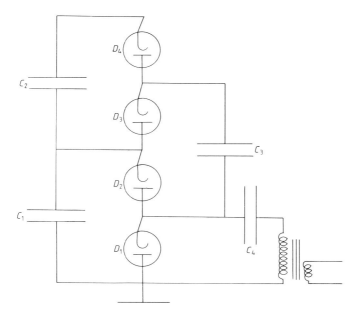

Figure 1.5.1 The voltage multiplying circuit devised by Cockcroft.

pump at earth potential. They, as well as the accelerating tube, were made out of straight glass cylinders, these being found to withstand high voltages much better than the largest glass bulbs obtainable.

Figure 1.5.2 is a photograph of the high-voltage equipment at the Cavendish Laboratory as it appeared towards the end of 1931. The tower of four rectifiers is on the left and the two-section accelerating tube is on the centre right of the picture. They were evacuated by separate oil diffusion pumps of the type which had recently been developed by C R Burch at Metropolitan-Vickers. Their use simplified greatly the problem of maintaining a sufficiently low pressure in the apparatus. High-vacuum technique was also made much simpler by the use of Plasticine and later by the use of Apiezon Sealing Compound Q. The pieces to be joined had merely to be placed together and the joint made vacuum-tight by pressing the compound with our fingers.

The proton source was of the type used by Aston, suitable power supplies being obtained from the belt-driven equipment on top of the tall white porcelain cylinder.

By April 1932 the apparatus would produce streams of protons accelerated by about 700 000 V and the day came when we could install a target for bombardment. A light element would obviously make the most suitable target and lithium seemed the most promising. We planned to detect any particles which might be ejected in a nuclear disintegration by the standard method of those days, namely a special microscope focused on a thin layer

Figure 1.5.2 The Cockcroft–Walton accelerator. Courtesy Cavendish Laboratory, University of Cambridge.

of finely ground willemite. Any fast particles would produce tiny flashes of light. The microscope was in a very small hut which we had built below the apparatus as a place of safety from high voltages and x-rays. I was carrying out the usual daily conditioning of the apparatus and happened to be alone in the room when the voltage reached its full value. I could not resist the temptation to leave the control table and have a look through the microscope. So I crossed the room to our hut in a crouching manner to avoid danger from high voltages. In the microscope there was a wonderful sight—lots of scintillations looking just like stars flashing out momentarily on a clear dark night. Very quickly we applied some simple checks to make sure that the scintillations represented genuine disintegrations. The results were then reported to Rutherford who very soon came along to have a look for himself. We manoevred him into our tiny hut and he too saw the scintillations and pronounced them to be produced by genuine alpha-particles. He said that *he* ought to know one when he saw it for he had been present at the birth of the alpha-particle and had been observing them ever since. So he had, and indeed he might have added that he had also christened them.

The work done in the next few days all supported our first ideas and the nuclear reaction occurring was obvious. A proton entered a lithium nucleus which then exploded into two alpha-particles travelling at high speed in opposite directions. Somewhat similar results could be obtained with other elements. Coming after so long a period of work with so little to show for it, it was most satisfying and exciting to get positive results bringing to light nuclear reactions not previously known. A new field had been opened up and the prospects for rewarding work in it were bright.

Nineteen thirty two, or thereabout, was an *annus mirabilis* for atomic physics. In 1919 Rutherford had shown how some elements could be transmuted by bombarding them with natural alpha-particles. Only one type of reaction was known—an alpha-particle went in and a proton came out. Considerable information about this type of reaction had been obtained but the practical limits of what could be done by existing techniques had almost been reached and the work was tending to become dull—hence the attraction of the new field opened up by the use of artificially accelerated particles. But other exciting things happened at the same time. Heavy hydrogen was discovered and this gave a new and important bombarding particle. The discovery of the neutron by Chadwick introduced a new nuclear particle which, having no electrical charge, was ideal for disintegration work. Artificial radioactivity and the positive electron [positron] were two further important discoveries at this time. I can well remember Rutherford's remark about it all—'It never rains but it pours'. He had been the leader in radioactive studies since the beginning and was greatly pleased that workers in his own laboratory were playing a large part in three of the new discoveries. It was a most exciting time to be there and one cannot but be grateful for the opportunity of experiencing it.

1.6 Cloud Chamber Researches in Nuclear Physics and Cosmic Radiation†

Patrick Blackett

The experimental researches with which I have been occupied during the 24 years of my career as a physicist have been mainly concerned with the use of Wilson's cloud chamber for the purpose of learning more about the intimate processes of interaction of the subatomic particles. On 12 December 1926, C T R Wilson gave his Nobel Lecture entitled 'On the cloud method of making visible ions and the tracks of ionising particles', and described in it how, after a long series of researches starting in 1895, he developed in 1912 this exquisite physical method. C T R Wilson was originally drawn to investigate the condensation of water drops in moist air through the experience of watching the 'wonderful optical phenomena shown when the sun shone on the clouds' surrounding his Scottish hilltops. I, like all the other workers with the cloud chamber the world over, are indebted more than we can express to his shy but enduring genius.

In 1919, Sir Ernest Rutherford made one of his (very numerous) epoch-making discoveries. He found that the nuclei of certain light elements, of which nitrogen was a conspicuous example, could be disintegrated by the impact of fast alpha-particles from radioactive sources, and in the process very fast protons were emitted. What actually happened during the collision between the alpha-particle and the nitrogen nucleus could not, however, be determined by the scintillation method then in use. What was more natural than for Rutherford to look to the Wilson cloud method to reveal the finer details of this newly discovered process. The research worker chosen to carry out this work was a Japanese physicist Shimizu, then working at the Cavendish Laboratory, Cambridge, to which Rutherford had recently migrated from Manchester. Shimizu built a small cloud chamber and camera to take a large number of photographs of the tracks of alpha-particles in nitrogen with the hope of finding some showing the rare disintegration processes. Unfortunately Shimizu had to return unexpectedly to Japan with the work hardly started. Rutherford's choice of someone to continue Shim-

† Extract from Blackett's Nobel Prize Speech of 1948, originally published by Elsevier.

izu's work fell on me—then in 1921 a newly graduated student of physics. Provided by Rutherford with so fine a problem, by C T R Wilson with so powerful a method, and by Nature with a liking for mechanical gadgets, I fell with a will to the problem of photographing some half million alpha-ray tracks.

In the autumn of 1931 in collaboration with G P S Occhialini, I started to study the energetic particles found in cosmic rays by means of the cloud method. About 4 years previously Skobeltzyn in Leningrad had investigated the beta rays from radioactive sources using a cloud chamber in a magnetic field of 1500 gauss. On some of the photographs he noticed a few tracks with very little curvature, indicating an energy over 20 MeV, that is much higher than any known beta ray. He identified these tracks with the particles responsible for the 'Ultrastrahlung' or 'cosmic rays', whose origin outside the earth's atmosphere had first been demonstrated in 1912 by the balloon flights of Hess and which had subsequently been much studied with ionisation chambers by Millikan, Kolhörster, Regener, Hoffman and others.

Skobeltzyn noticed also that these energetic particles occasionally occurred in small groups of 2, 3 or 4 rays, apparently diverging from a point somewhere near the chamber.

Skobeltzyn's work was followed up by Kunze in Kiel, and by Anderson in Pasadena. By using much larger magnetic fields up to 18 000 gauss, the energy spectrum of the particles was shown by these workers to extend to at least 5000 MeV, and it was found that roughly half the particles were positively, and half negatively charged. The occasional association of particles was again noticed, particularly by Anderson.

The method used, that of making an expansion of a cloud chamber at a random time and taking the chance that one of the rare cosmic rays would cross the chamber during the short time of sensitivity—generally less than ¼ second—was much consuming of time and photographic film, since in a small chamber only some 2–5% of photographs showed cosmic ray tracks.

Occhialini and I therefore set about devising a method of making cosmic rays take their own photographs, using the recently developed Geiger–Müller counters as detectors of the rays.

Bothe and Rossi had shown that two Geiger counters placed near each other gave a considerable number of simultaneous discharges, called coincidences, which indicated in general the passage of a single cosmic ray through both counters. Rossi devised a neat valve circuit by which such coincidences could easily be recorded.

Occhialini and I decided to place Geiger counters above and below a vertical cloud chamber, so that any ray passing through the two counters would also pass through the chamber. By a relay mechanism, the electric impulse from the coincident discharge of the counter was made to actuate the expansion of the cloud chamber, which was made so rapid that the ions produced by the ray had no time to diffuse much before the expansion was complete. The chamber was placed in a water-cooled solenoid giving 3000 gauss. Having made the apparatus ready, one waited for a cosmic ray to arrive and take its own photograph. Instead of a small fraction of photographs showing a cosmic ray track, as when using the method of random expansion, the counter-controlled chamber yielded a cosmic ray track on 80% of the photographs. The first photographs by this new method were made in the early summer of 1932.

In the autumn of the same year, Anderson working with a normal chamber taking photographs at random, reported the finding of a track which he interpreted as showing the existence of a new particle—the positive electron.

The track described by Anderson traversed a lead plate in the centre of the chamber and revealed the direction of motion of the particle by the difference of curvature on the two sides. From the direction of motion and the direction of the magnetic field, the charge was proved positive. From the range and ionisation, the mass could be proved to be much less than that of a proton. Anderson thus identified it as a new particle, the positive electron or positron.

During the late autumn of 1932, Occhialini and I, using our new counter-controlled cloud method, accumulated some 700 photographs of cosmic rays, among which groups of associated rays were so striking a feature as to constitute a new phenomenon and to deserve a name. From their appearance they came be to known as 'showers' of cosmic ray particles. As many as 23 particles were found on a single photograph, diverging from a region over the chamber. Roughly half the rays were due to positively charged and half to negatively charged particles. From their ionisation and range, the masses of the positive particles was evidently not much different from that of negative electrons. So not only was Anderson's discovery of the positive electron further confirmed by a wealth of evidence, but it was proved that the newly discovered particles occurred mainly in showers along with an approximately equal number of negative electrons. This fact of the rough equality of numbers of positive and negative electrons, and the certainty that the former do not exist as a normal constituent of matter on the earth, led us inevitably to conclude that the positive electrons were born together in collision processes initiated by high-energy cosmic rays. The energy required to produce such a pair is found from Einstein's famous equation to be $2 mc^2 \simeq 1$ MeV. So was demonstrated experimentally for the first time the transformation of radiation into matter.

The fate of the positrons was discussed in relation to Dirac's theory of holes. On this theory a positive electron was envisaged as a 'hole' in a sea consisting of an infinite number of negative electrons in states of negative kinetic energy. Dirac's theory predicted that a positive electron would disappear by uniting with a negative electron to form one or more quanta. Occhialini and I suggested that the anomalous absorption of hard gamma rays by nuclei might be a result of the process of pair production, and that

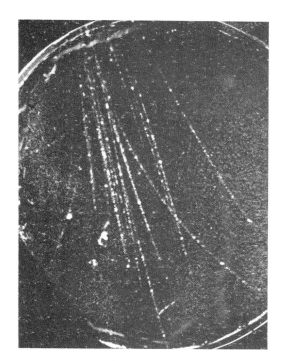

Figure 1.6.1 Cosmic ray showers, photographed by Blackett and Occhialini. *Left*: One of the first photographs of a large shower of cosmic ray particles. Some 16 particles, about half positive and half negative, diverge from a region over the chamber. This shower was interpreted as showing the birth of a number of pairs of positive and negative electrons. The counter-controlled cloud chamber was in a field of 3000 G. *Below*: Some 23 particles cross the chamber. Several radiant points can be detected above the chamber and also in the lead plate. H = 2000 G. From *Proc. R. Soc.*. A **139** 721, 723 (1933).

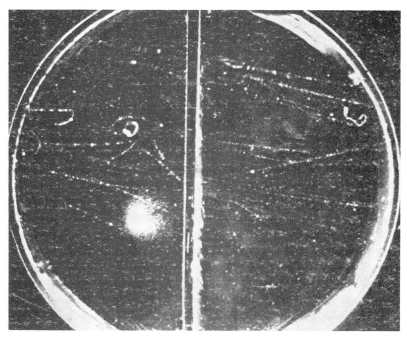

the observed re-emission of softer radiation might represent the emission of two 0.5 MeV quanta resulting from the annihilation of a positive and negative electron. Subsequent work has confirmed this suggestion.

This work was described in a paper which appeared in March 1933. Some of the photographs from the paper are reproduced here [figure 1.6.1]. These represent the first published photographs showing positive electrons, as Anderson's very beautiful photograph, though taken six months earlier, was not published till shortly afterwards.

The photographs showed clearly that some form of non-ionising radiation must play an essential part in the formation of the showers, and that the mean range in lead of these radiations, which were assumed to be either photons or neutrons, must be quite small. Subsequent theoretical work by Heitler, Bethe, Bhabha and others gave a full account of these showers as due to a cascade process, consisting of the alternate emission of collision radiation by fast electrons and positrons, and the subsequent absorption of the latter by pair production.

As soon as the presence of positive electrons in cosmic rays was fully established, experiments were undertaken in collaboration with Occhialini and Chadwick to see if they were formed when hard gamma rays from radioactive sources were absorbed by matter. This was found to be the case when the energy of the rays was considerably above 1 MeV.

It is interesting to note that the development of the counter-controlled cloud chamber method not only attained the original objective of achieving much economy in both time and film, but also proved to have the quite unexpected advantage of greatly enhancing the number of associated rays photographed. This was so because the greater the number of rays in a shower of cosmic ray particles, the greater the chance that the counter system controlling the chamber would be set off. As a result the larger showers appeared in the photographs far more frequently relative to single rays than they actually occur in nature. This property of bias towards complex and so interesting phenomena has proved one of the most important advantages of the counter-controlled method.

In a subsequent paper I sketched in detail the formation of tracks by the counter-controlled method and calculated the expected breadth of a track as a function of the coefficient of diffusion of the gaseous ions and of the time elapsing between the passage of the rays and the completion of the expansion. The experimentally measured breadths in hydrogen and oxygen agreed well with the theory.

1.7　Blackett and the Positron

Paul Dirac

One of the central figures in the Cavendish during the period around 1930 was P M S Blackett. He specialised in developing the Wilson chamber. The photographs were mostly taken in a magnetic field, so that the tracks were curved.

I was quite intimate with Blackett at the time and had told him about my relativistic theory of the electron. According to this theory there were negative-energy states for an electron, which were normally filled up, with one electron in each state. But there could be an empty negative-energy state. This would appear as a new kind of particle with a positive charge and, as was pointed out by several people, with the same mass as the electron. Such a particle was afterwards named a positron.

I discussed this theory with Blackett and we wondered whether the theory was correct and whether positrons really existed. A positron in a Wilson chamber would give a track like an electron, but it would be curved in the opposite direction under the influence of the magnetic field. If one examined a particular track one could not tell if it represented an electron or a positron, because one did not know in which way the particle that had produced it had been moving.

In looking over many of Blackett's photographs and assuming a likely direction for the motion of the particle from the circumstances of the experiment, one seemed to have plenty of evidence for positrons. But one could not be sure, and Blackett would not publish such uncertain evidence.

Then Blackett noticed that, if he had a radioactive source in the Wilson chamber, many of the particles coming out from it had tracks curved to correspond to positrons. This seemed to me to be pretty conclusive evidence. But Blackett was not satisfied. He argued that there might be electrons from outside which, by chance, ran into the source. This was most unlikely, but not impossible. So Blackett still would not publish his findings.

In order to settle the question Blackett proceeded to work out the statistics of how many chance electrons would have to run back into the source to account for his observations, and see if it was at all reasonable. While

Blackett was engaged in this work the news came that Anderson had discovered the positron.

Anderson's work involved just a single track, which passed through a lead plate, and was more strongly curved on one side than the other. The particle was moving more slowly where the track was strongly curved. The particle could only lose energy in passing through the plate, so its direction of motion was fixed.

If Blackett had been less cautious, he could have been first in publishing evidence for the positron. Anderson knew nothing about my theory. In retrospect, it was found that for several years previously people had been obtaining pictures of radioactive sources emitting positrons without realising what they were. Some had even been published.

Part 2

Cambridge Physics and the Cavendish

2.1 Introduction

Given a run of achievements such as that described in Part 1 it is natural to ask about the environment that fostered it. In this part we therefore take a general look at the Cavendish Laboratory, and at the rest of Cambridge physics, in the late 1920s and early 1930s.

Of course, this forms only one part of the physicists' environment, and to the outsider it might seem a small part. The Cambridge of the period was that of Russell and, at weekends at least, of Keynes, in which physics was only one of several areas of outstanding intellectual achievement and excitement. The period was also marked by major social and political developments, with a deepening depression in Britain and the rise of Hitler and Mussolini in Europe. We shall be returning to the latter events in Part 4 of this volume, as the rise of Hitler in particular was to have a profound effect upon the lives of physicists and upon the geography of physics. But the most striking observation in the present context is that the written recollections of the physicists barely touch upon the world outside physics. A few of the contributors to this volume were asked explicitly to describe the social as well as the scientific environment of the early 1930s in Cambridge, but all preferred to stick to science and it is clear, and perhaps natural given the scientific excitement of the era, that this was what dominated their lives. They were of course human. They had their families and their colleges (of which the latter in particular could make very heavy demands on time), their sports and their hobbies; but even here the recorded memories barely break away from science.

Several members of the Cavendish played golf, and Ralph Fowler in particular, one of the honorary members of the laboratory, was quite distinguished at the sport. However they always played together. The regular members of the Sunday foursome, all Trinity men, were Fowler, Aston, Rutherford and G I Taylor, and the last-named recalled that they went out not so much for the golf as to hear Rutherford talk.[1] Taylor was also a keen sailor, and owned a 19 ton, 48 foot cutter named *Frolic*, in which he and his wife Stephanie explored the islands of Scotland and Norway. Despite the name of the boat it remains doubtful whether he ever got away from physics. His speciality was, after all, fluid mechanics, and he often travelled with a fellow scientist, George McKerrow of the Metropolitan-Vickers Company. In the 1930s he set up a company with McKerrow and another colleague to manufacture a new design of anchor he had invented in the course of his sailing.[2] C T R Wilson climbed mountains, but once there seems to have concentrated as much on the science of cloud formation as on the view. Cockcroft, Blackett and Massey all played hockey for, of course, a laboratory team.

Figure 2.1.1 The golfing foursome, photographed at a British Association meeting in South Africa, 1929. Left to right: R H Fowler, F W Aston, Rutherford and G I Taylor. Reproduced by kind permission of Sir Mark Oliphant from his book *Rutherford: Recollections of the Cambridge Days* (Amsterdam: Elsevier, 1972).

Of the social life of the physicists there are few glimpses, except perhaps for the tales of Rutherford's weekend tea parties.[3] In the middle of all the scientific excitement of 1932, however, Norman Feather found time for courtship and marriage, Patrick Blackett found time for politics, and Peter Kapitza, in 1933, found time perforce to argue the merits of modern art in the face of conservative uproar over the portrait of Rutherford commissioned from Eric Gill for the new Mond Laboratory.[4] Even in this case, however, the authority brought in to secure victory was none other than the physicist Niels Bohr. It is to physics that we shall devote the rest of this introduction.

The location of the famous experiments of 1932 was the Cavendish Laboratory, which, with its 'Nursery' and 'Garage' is described in the course of the recollections below. It was, as John Ratcliffe has recalled, a 'dingy, dirty and dismal' sort of place,[5] built in the nineteenth century in a style that caused no problems until the mid-1920s but was not then easily adapted to accommodate the advance of physics. On more than one occasion in the twenties and thirties the entrances to the rooms and even that to the labora-

RUTHERFORD

Figure 2.1.2 Rutherford by Eric Gill. Reproduced by kind permission of Sir Mark Oliphant from his book *Rutherford: Recollections of the Cambridge Days* (Amsterdam: Elsevier, 1972).

tory itself through the arch from Free School Lane were the limiting factors on the design of the large-scale electrical apparatus being introduced. Inside, everything was improvised. The budget was minimal and expenditure was very tightly controlled. Round about 1932 the electricity supply was standardised, having consisted until then of a 200 volt AC supply at 93 cycles per second from the city and a 200 volt DC supply from the laboratory's own generators. This did not however affect the workshop, for most of the machines were still hand-driven. In general the physicists had to make all their apparatus for themselves, usually out of whatever bits and pieces they could find or scrounge, and the result was inevitably something of a mess.[6]

In the midst of all the mess, two men ruled the show. One was the workshop supervisor and stores keeper, Lincoln, who measured out electric cable by the inch and rationed the supply of screws. The other was Rutherford. Rutherford was a large man with a booming voice, an enormous heart, a wicked sense of humour, and one of the most perceptive physical minds ever. He was scientific genius and humanity together writ large. The year 1932 found him recently raised to the peerage but working as hard as ever, following up the innovations and achievements of his staff. Recent developments of electronic instrumentation by Ward and Wynn-Williams had opened up new possibilities for the accurate analysis of the spectra of radioactive emissions, and in 1932 Rutherford, working with Wynn-Williams, Lewis and Bowden (now Lord Bowden), was in the thick of the exploration of these spectra.[7] Following Cockcroft and Walton's disintegration of lithium, he also launched himself into their project, working first on their original apparatus and then with a set-up of his own.[8]

Rutherford dominated the Cavendish both physically and intellectually, but there were also a number of other professors occupying a variety of positions *vis-a-vis* the laboratory. Sir J J Thomson, Rutherford's predecessor as director of the laboratory and then Master of Trinity College, continued to conduct experiments in the 'Garage', on what would now be called plasma oscillations in a toroidal electrodeless discharge. Ralph Fowler, appointed in 1932 to a Chair in the Faculty of Mathematics, and Geoffrey Taylor, Yarrow Research Professor of the Royal Society, both had rooms in the Cavendish though not formally members of its staff, while C T R Wilson, whose Chair was at the Cavendish, preferred to work in the peace and quiet of the Solar Physics Observatory. C T R, then in his sixties, was still developing the cloud chamber that bears his name, and still pursuing his study of thunderstorms. Fowler and Taylor, who were originally mathematicians, had both reacted strongly to Rutherford's arrival in Cambridge and Trinity. Both were greatly inspired by his presence, and in turn both found themselves loved and respected by Rutherford. Taylor, whose Royal Society professorship was a non-teaching one (he was paid, as Rutherford put it, providing he did no work) conducted his experimental research on fluids in a room on the main floor of the laboratory, where Fowler also had an office. Both covered a prodigious range in their research. Taylor was one of very few scientists in the present century to have shown equal brilliance in both mathematics and experiment, and he had made the science of fluid dynamics, in all its facets, his own. In 1932 he was working mainly on the theory of turbulence, but also on the viscosity of mixed fluids and a number of other subjects.[9] Fowler's range covered pure mathematics, statistical mechanics (for which he is best known today), astrophysics (in which he did some of his most original research), quantum theory and thermodynamics. In 1932 he was working on a variety of topics, including the theory of semiconductors with Alan Wilson,[10] and the quantum mechanics of electrochemistry with R W Gurney.[11]

Apart from the professors, the Cavendish also had several other most distinguished members. There was Peter Kapitza, who himself received a Royal Society professorship in 1933 when his new Mond Laboratory was opened, and who spent 1932 preparing for this opening and continuing his pioneering work on high magnetic fields.[12] F W Aston was continuing his work on mass spectroscopy, and in 1932 he published details of the isotopes and atomic weights of a wide range of medium weight elements.[13] Charlie Ellis was studying x-ray and gamma-ray spectra, and collaborating with the young mathematician Nevill Mott in trying to interpret his results.[14] Patrick Blackett worked with Occhialini on the automatic cloud chamber, and with D S Lees on more conventional cloud chamber observations.[15] James Chadwick, who was Rutherford's assistant director of research, supervised the activities of the many research students and worked first on the discovery and then on the properties of the neutron. John Cockcroft, while supervising the building and equipment of the Mond Laboratory for Kapitza, found time to disintegrate a few atoms. John Ratcliffe, who had taken over Appleton's laboratory in the late twenties, pursued his radio investigations of the ionosphere.[16]

These were the big names in the Cavendish, and they alone constitute a remarkable array of talent, but in the Cambridge physics community of that period they were only the tip of an iceberg. Rutherford's reputation ensured that the laboratory was also full of brilliant younger men, among whom those we have already encountered, Walton, Feather, Dee and Occhialini, were only a few. There was the brilliantly inventive Wynn-Williams, who during the 1931–2 session devised and constructed the scale-of-two counter.[17] And there was Mark Oliphant (now Sir Mark), who worked with Philip Moon, R C Evans and R M Chaudhri on surface ionisation phenomena.[18] Harrie Massey (now Sir Harrie), C B Mohr, F H Nicoll and J McDougall studied electron scattering, while Massey also pursued his theoretical studies on nuclear collisions.[19] F C Webster, L H Gray, T G P Tarrant and C Y Chao all investigated the production and properties of hard gamma-rays, Gray and Tarrant completing their famous experiments on the annihilation energy produced when helium was bombarded with such rays.[20] F C Champion, Norman Alexander (now Sir Norman) and F R Terroux investigated the spectra of beta-ray emissions,[21] while J L Pawsey worked with Ratcliffe on the ionosphere,[22] and J K Roberts studied the heat exchange between gases and solids.[23] Of these young men many later became major figures in the world of nuclear physics, while others found fame elsewhere, such as Louis Gray in radiobiology and Pawsey in radioastronomy.[24]

Still the talent is not exhausted, for physics was also pursued outside the Cavendish, and especially within the Mathematics Faculty.

Apart from Fowler there were also the young Paul Dirac, recently appointed Lucasian Professor in succession to Sir Joseph Larmor and, before him, Stokes and Newton; and Sir Arthur Stanley Eddington, Plumian Professor and director of the Observatory. Both were world famous, Dirac for his quantum theory of the electron and quantum electrodynamics, and Eddington for his contributions to astrophysics and relativity theory. Dirac

Figure 2.1.3 Cavendish physics research students, 1932. Courtesy Cavendish Laboratory, University of Cambridge.

had recently published his theory of the magnetic monopole and was at the time continuing his study of relativistic quantum mechanics.[25] Towards the end of 1932 he also embarked on a joint project with Kapitza, analysing the reflection of electrons from standing waves.[26] Eddington had recently published his book on the expanding universe and was still working on astrophysics, especially on the densities of stars.[27] But his main research was also on relativistic quantum mechanics, as he sought to derive the proton and electron masses and other fundamental constants of nature from basic theoretical considerations.[28]

Others formally classed as mathematicians included Alan Wilson (now Sir Alan) who had recently provided the foundations for the quantum theory of semiconductors and was then working on the theory of metals.[29] Nevill Mott (now Sir Nevill) was working on electron polarisation and other subjects closely related to the experimental work of the Cavendish.[30] Subrahmanyan Chandrasekhar was commencing his research in astrophysics.[31] R W Gurney studied the quantum mechanics of electrochemistry,[32] while H M Taylor investigated the theory of anomalous alpha-particle scattering.[33] Other theoretical physicists active at the time included J M Jackson, H R Hulme and the alphabetic J K L MacDonald, as well as Harold Jeffreys (now Sir Harold), whose work in 1932 included a study of plasticity and creep in solids.[34]

Also associated with the Cavendish at this time were E C Bullard, in the Department of Geodesy and Geophysics, and the members of the crystallography side of the Department of Mineralogy—J D Bernal, C P Snow, F I G Rawlins and W A Wooster. The crystallographers had dissociated from the mineralogists in 1931 and though geographically unmoved they now came under the auspices of the Cavendish; but this change had made little difference to their close relationship with the laboratory. In 1932 Bernal, already a world authority on his subject, began a joint research programme with Ralph Fowler on the quantum theory of the H_2O molecule.[35] Wooster had originally moved from the Cavendish, with which he retained close connections, while Snow, later to find fame as a writer rather than as a scientist but then working with Rawlins on modified ionic states in crystals,[36] was also a Cavendish familiar. Finally, mention should also be made of the Colonel, Professor 'Chubby' Stratton, head of the Solar Observatory where C T R Wilson preferred to work. Stratton had recently fallen seriously ill and was not to get back to full working health for another couple of years, but he remained general secretary of the British Association for the Advancement of Science, and he was also active in his college, Gonville and Caius, where also were Chadwick and several other of the physicists.[37]

The above catalogue may serve as an introduction to the galaxy of talent that was Cambridge physics in 1932. In the reminiscences that follow, as well as in the rest of those in this volume, may be found deeper and more personal recollections of the stars, and of the ambience in which they moved.

Notes

1   The material in this section is taken largely from the biographical memoirs listed in the bibliography. For Taylor see *Biog. Mem. Fell. R. Soc.* **22** 565 (1976) and also J G Crowther *The Cavendish Laboratory 1874–1974* (London: Macmillan, 1974) p. 249

2   1976 *Biog. Mem. Fell. R. Soc.* **22** 565

3   See for example Oliphant M L 1972 *Rutherford: Recollections of the Cambridge Days* (Amsterdam: Elsevier) p. 21

4   Oliphant M L 1972 *Rutherford: Recollections of the Cambridge Days* (Amsterdam: Elsevier) p. 92

5   Lecture delivered at the conference to commemorate the 50th anniversary of the discovery of the neutron. Conference proceedings published in *The Neutron and its Applications 1982* (Inst. Phys. Conf. Ser. 64) (Bristol: Institute of Physics, 1983)

6   For a description of the laboratory see, for example, Oliphant's book cited in note 3

7   Lord Rutherford and Bowden B V 1932 *Proc. R. Soc.* A **136** 407; Lord Rutherford, Wynn-Williams C E, Lewis W B and Bowden B V 1933 *Proc. R. Soc.* A **139** 617; see also Lewis W B and Wynn-Williams C E 1932 *Proc. R. Soc.* A **136** 349

8   See the paper by Oliphant in Part 4 below and the notebooks of Cockcroft and Walton, Churchill College, Cambridge.

9   For example Taylor G I 1932 *Proc. R. Soc.* A **135** 678, 685; Taylor G I 1932 *Mem. R. Meteorol. Soc.* **4** no 35 43; Taylor G I and MacColl J W 1933 *Proc. R. Soc.* A **139** 287, 298; Taylor G I 1932 *Proc. R. Soc.* A **138** 41

10 For example Fowler R H 1933 Proc. R. Soc. A **140** 505; Fowler R H 1932 *Trans. Faraday Soc.* **28** 368; Fowler R H and Wilson A H 1932 *Proc. R. Soc.* A **137** 503

11 For example Fowler R H and Gurney R W 1931 *Proc. R. Soc.* A **136** 378. Fowler also worked with Bernal, for which see below.

12 For example Kapitza P 1932 *Proc. R. Soc.* A **135** 537

13 For example Aston F W 1932 *Proc. R. Soc.* A **134** 554

14 For example Ellis C D 1932 *Proc. R. Soc.* A **136** 396; Ellis C D 1932 *Proc. R. Soc.* A **138** 318; Ellis C D 1932 *Proc. R. Soc.* A 1932 **139** 336; Ellis C D and Mott N F 1932 *Proc. R. Soc.* A **139** 364

15 For example Blackett P M S and Occhialini G 1932 *Nature* **130** 363; Blackett P M S and Occhialini G 1933 *Proc. R. Soc.* A **139** 699; Blackett P M S and Lees D S 1932 *Proc. R. Soc.* A **134** 658; Blackett P M S and Lees D S 1932 *Proc. R. Soc.* A **136** 325, 338

16 For example Ratcliffe J A and Pawsey J L 1933 *Proc. Camb. Phil. Soc.* **29** 301

17 Wynn-Williams C E 1932 *Proc. R. Soc.* A **136** 312

18 For example Moon P B 1932 *Proc. Camb. Phil. Soc.* **28** 490; Chaudhri

R M 1932 *Proc. Camb. Phil. Soc.* **28** 349; Evans R C 1932–3 *Proc. Camb. Phil. Soc.* **29** 161, 522; Evans R C 1933 *Proc. R. Soc.* A **139** 604; Moon P B and Oliphant M L 1932 *Proc. R. Soc.* A **137** 463; Chaudhri R M and Oliphant M L 1932 *Proc. R. Soc.* A **137** 662

19   For example Massey H W and Mohr C B 1932 *Proc. R. Soc.* A **136** 289; Massey H W and Mohr C B 1933 *Proc. R. Soc.* A **139** 187; Mohr C B and Nicoll F H 1932 *Proc. R. Soc.* A **138** 229, 469; McDougall J 1932 *Proc. R. Soc.* A **136** 549; McDougall J 1932 *Proc. R. Soc.* A **138** 550; McDougall J 1932 *Proc. Camb. Phil. Soc.* **28** 341; Massey H W 1932 *Proc. Camb. Phil. Soc.* **28** 99; Massey H W 1932 *Proc. R. Soc.* A **137** 447; Massey H W 1932 *Proc. R. Soc.* A **138** 460

20   For example Chao C W 1932 *Proc. R. Soc.* A **135** 206; Tarrant T G P 1932 *Proc. R. Soc.* A **135** 223; Gray L H and Tarrant T G P 1932 *Proc. R. Soc.* A **136** 662; Gray L H and Tarrant T G P 1932 *Proc. Camb. Phil. Soc.* **28** 124; Webster H C 1932 *Proc. R. Soc.* A **136** 428

21   For example Champion F C 1932 *Proc. R. Soc.* A **136** 630; Champion F C 1932 *Proc. R. Soc.* A **137** 688; Alexander N S and Terroux F R 1932 *Proc. Camb. Phil. Soc.* **28** 115

22   See note 16 above

23   For example Roberts J K 1932 *Proc. R. Soc.* A **135** 192

24   1964 *Biog. Mem. Fell. R. Soc.* **10** 229; 1966 *Biog. Mem. Fell. R. Soc.* **11** 195

25   Dirac P A M 1931 *Proc. R. Soc.* A **133** 60; Dirac P A M 1932 *Proc. R. Soc.* A **136** 453

26   Dirac P A M and Kapitza P 1933 *Proc. Camb. Phil. Soc.* **29** 297

27   Eddington A S 1932 *The Expanding Universe* (Cambridge: Cambridge University Press); Eddington A S 1932 *Mon. Not. R. Astron. Soc.* **92** 364, 471; Eddington A S 1933 *Mon. Not. R. Astron. Soc.* **93** 320

28   For example Eddington A S 1931 *Proc. R. Soc.* A **134** 524; Eddington A S 1932 *Proc. R. Soc.* A **138** 17

29   Wilson A H 1931 *Proc. R. Soc.* A **134** 277, 458; Wilson A H 1932 *Proc. R. Soc.* A **136** 487; Wilson A H 1932 *Proc. R. Soc.* A **138** 594

30   For example Mott N F 1932 *Proc. R. Soc.* A **135** 429; Jackson J M and Mott N F 1932 *Proc. R. Soc.* A **137** 703; Taylor H M and Mott N F 1932 *Proc. R. Soc.* A **138** 665; Ellis C D and Mott N F 1933 *Proc. R. Soc.* A **139** 364

31   For example Chandrasekhar S 1932 *Proc. R. Soc.* A **135** 472

32   For example Gurney R W 1932 *Proc. R. Soc.* A **136** 378

33   For example Taylor H M 1932 *Proc. R. Soc.* A **136** 605

34   Jeffreys H 1932 *Proc. R. Soc.* A **138** 283

35   For example Bernal J D and Fowler R H 1933 *J. Chem. Phys.* **1** 515; Bernal J D and Fowler R H 1933 *Trans. Faraday Soc.* **29** 1049

36   Snow C P and Rawlins F I G 1932 *Proc. Camb. Phil. Soc.* **28** 522

37   1961 *Biog. Mem. Fell. R. Soc.* **7** 281

2.2   Some Recollections of Low Energy Nuclear Physics†

John Cockcroft

I started work as a student in the Cavendish Laboratory in 1922 towards the end of what might be described by the irreverent as the Old Stone Age of nuclear physics. In the ground floor research room, designed and once used by Clerk Maxwell, Rutherford worked with Chadwick and his research assistant, George Crowe. Their studies on the disintegration by alpha-particles of boron, nitrogen, fluorine, aluminium and phosphorus were carried out with a 30 millicurie Radium C source of alpha-particles and a zinc sulphide screen as a detector of the disintegration protons. After sitting in the darkened room for half an hour or so they could count up to about 60 of the faint proton scintillations a minute before getting confused by simultaneous scintillations. Crowe has given a description of Rutherford's methods. 'He was shooting protons out of light atoms by means of alpha-particles; carbon would not play. The next day was the turn of aluminium and on some theoretical grounds Rutherford thought that protons from aluminium would have a high velocity and a long range. "Now, Crowe, have some mica absorbers ready with a stopping power of 50 cm." "Yes Sir." "Now, Crowe, put in a 50 cm screen." "Yes Sir". "Why don't you do what I tell you; put in a 50 cm screen." "I have Sir." "Put in 20 more." "Yes Sir." "Why the devil don't you do what I tell you. I said 20 more." "I did Sir." "There's some damned contamination; put in two 50s." "Yes Sir." "Ah, it's all right that's stopped 'em. Crowe my boy you're always wrong until I've proved you right. Now we'll find the exact range!" '

In another less ancient part of the laboratory Blackett introduced more advanced techniques by building his cloud chamber for the study of the nitrogen disintegration and was rewarded by obtaining his classical pictures giving the first direct proof that the alpha-particle was captured in the disintegration. From this Blackett moved on to construct his Geiger counter

† This paper is an extract from one presented at an International Conference on Fast Neutron Physics, Rice University, Houston, Texas, February 1963. Originally published in a volume devoted to that conference by the Rice University.

controlled cloud chambers for the study of the cosmic radiation and this led in turn to his work with Occhialini on the production of positron and negative electron pairs, following immediately after Carl Anderson's discovery of the positron. I remember Millikan telling us about Anderson's discovery in the Cavendish Physical Society Meeting and the irreverent P I Dee getting up to point out that one of the tracks was curved in both positive and negative directions. However, it did turn out that Carl Anderson was right.

Rutherford's Manchester interests in the scattering of alpha-particles by nuclei were continued using the same techniques by Chadwick and Bieler and later on by Chadwick and Rutherford. Marked divergence from Rutherford's scattering was observed in scattering of alpha-particles from light nuclei and attempts were made to explain the divergencies by invoking magnetic forces or a plate-like structure of the alpha-particle.

In the upper region of the laboratory Ellis measured the energies of the gamma rays from the heavy elements using magnetic focusing of the electrons and this was correlated with the fine structure of alpha-particles and the long range alpha-particles.

In 1930 a new line of work was triggered off in the laboratory by Bothe and Becker's discovery that the bombardment of beryllium by polonium alpha-particles led to the emission of radiation which was more penetrating than any known gamma rays. With characteristic speed of response H C Webster was put on to study the radiation and concluded that the energy of the gamma radiation from beryllium would be about 7 MeV and the boron radiation about 10 MeV. The radiation was then put into a cloud chamber but no unusual effects were observed—the sources were later seen to have been too weak. However, soon after this the Joliot-Curie observation that the radiation could project hydrogen nuclei with energies requiring 50 MeV gammas led Chadwick to a high pressure investigation in which the atmosphere of secrecy across the corridor from us was only paralleled by the later secrecy of the Manhatten Project.

This was the environment in which from the autumn of 1924 I learnt my nuclear physics. I was put to work as an apprentice in the laboratory attic by Chadwick together with other embryo nuclear physicists. We had to practise alpha-particle and proton counting using scintillation screens and measured our efficiencies by dual observer counting methods. We built gold leaf electroscopes to measure gamma ray intensities. We learnt how to produce what passed for high vacuum in those days, first using the Fleuss reciprocating hand pump for the fore-vacuum and experimenting with the first diffusion pump ever seen in the laboratory—a US make. We blew Macleod gauges to measure our vacua and I had the salutary experience of destroying one in the classical manner just as Rutherford walked in at the door whilst Joseph Boyce and I were recovering mercury from the channels between the floor boards. After serving my time for a few months I was diverted to make use of my electrical engineering training to help Kapitza to install his alternator, which he short circuited to produce magnetic fields of up to 300 000 gauss in coils designed on highly theoretical principles by me. I also attended the Tuesday evening meetings of the Kapitza Club presided over by Kapitza, where the latest important papers in the Cavendish Laboratory field were presented with very frequent interruptions by Kapitza.

Rutherford's Royal Society Presidential Address delivered on 30th November 1927, urged 'the development of sources of atoms and electrons with an energy far transcending that of alpha-particles and beta particles from radioactive matter'. The first response to this was the arrival of T E Allibone from Metropolitan-Vickers, bringing with him a 500 kV Tesla coil with rotating spark gap excitation which upset the radio sets within a quarter of a mile around. He installed this in our already crowded laboratory. To this he connected a home-built 300 kV electron tube and he produced intense beams of electrons, directing them on to fluorescent minerals for the edification of visitors like Rutherford, who must have received quite a few roentgens as he admiringly handled the crystals. Allibone stuck to electron work but his success in operating accelerating tubes at 300 kV was an important link in the chain of events which led me to take up Rutherford's challenge. Fortunately I was saved from attempting to rival the energies of the alpha-particles from radium compounds by the timely visit of Gamow in November, 1928, to expound his theory of the quantum theory of atomic disintegration which he had developed at Bohr's laboratory in Copenhagen— an example of the good effects of wandering scholars in spreading the gospel.

He showed how the disintegration of elements such as aluminium by alpha-particles from polonium could be explained by the quantum mechanical formulae for the penetration of alpha-particles through the nuclear potential barrier. This led me in turn to calculate the probability of protons of a few hundred kV energy penetrating the barriers of light nuclei. I calculated that 300 kV protons should produce about 2 million disintegrations per minute per microampere of protons impinging on boron. The result of this chain of events was that I decided to try to accelerate protons first by a steady potential of 300 kV and my notebooks record the initial playing about with the components of a high voltage, high vacuum system— the first oil diffusion pumps ever made were bought from C R Burch at Metropolitan-Vickers for 50 shillings. Tests were made on low vapour pressure vacuum stopcock grease and low vapour pressure Plasticine also produced by Burch.

I was then joined by E T S Walton who had at Rutherford's suggestion worked for a time on a primitive betatron and also on an equally primitive linear accelerator of a Sloan-Lawrence type.

By mid-1930 we had installed a 250 kV transformer and built a voltage doubling rectifier circuit for 300 kV and hooked this to an accelerating tube with a high voltage canal ray tube as a proton source. We then bombarded a wide range of targets from lithium to lead with several microamperes of protons and looked for gamma radiation which we hoped to find from proton capture but found only inhomogeneous radiation with a maximum energy of 40 kV.

At this point of our work in May 1931, when we ought to have been looking for alpha emission we had to move out of our laboratory and we then decided to build a new apparatus for 500–600 kV and for this we designed a voltage quadrupling circuit of a modified Greinacher type and checked this on a laboratory bench. Six months later we were operating again in our new research room. The main difficulty we had with this apparatus was in maintaining a high enough vacuum. The joints between the glass cylinders and the metal sheaths were made with a low vapour pressure Plasticine. We had to go round and round the joints with our thumbs to seal up the leaks spending hours on the job. Success was indicated by a discharge tube mounted on top of the first stage of our oil diffusion pumps showing green fluorescence and then a black-out. We outgassed the tubes by repeated discharges until they were hard. The DC voltage on the accelerating tube was measured by the spark length between two large aluminium spheres and checked by a magnetic deflection of the protons after acceleration—the highest energy being about 500 kV.

We wasted a certain amount of time in these running-in experiments and there was some impatience on Rutherford's part, but on 13 April we bombarded a lithium target with currents of about 2 microamperes and observed large numbers of bright scintillations on a zinc sulphide screen placed outside the apparatus. These were so bright that they were obviously alpha-particles and not protons. The number of scintillations increased 40-fold as the accelerating voltage increased from 126 to 250 kV and the range of the alpha-particles was found to be about 8.4 cm. There was a certain amount of fuss made by the Press after Rutherford made the announcement of this discovery at the Royal Society Meeting and Rutherford told the Press that the idea of obtaining power from nuclei was 'all moonshine'.

We soon found that the number of alpha-particles was so great that we had to abandon our primitive counting techniques and install a parallel plate ionisation chamber coupled to a linear amplifier and oscillograph and record the disintegration particles on photographic film which we took home laboriously to count the number of kicks. We were fortunately saved from further labours of that kind by the introduction of scalers due to the work of Wynn-Williams labouring in an underground cellar of the laboratory with no objective other than the love of new techniques characteristic of the laboratory at this time.

Another important addition to our techniques was provided by the diversion of P I Dee from his work with C T R Wilson to build a cloud chamber fed with disintegration products from a second accelerating tube in our laboratory. Dee had considerable difficulty in outgassing his tube, owing to lubricating oil having been left by the workshop in the screw threads of his ion source. When this difficulty was overcome, considerable difficulties of scheduling arose and it was necessary to arrive by 9.00 AM to establish priorities. By mid-1933 the apparatus was working well and Dee and Walton produced very fine photographs of the lithium and boron transmutations just after similar work had been published by Kirchner in Germany.

Rutherford's enthusiasm for the new horizons opened up led him to divert Oliphant from his work with positive ions to build a 200 kV accelerator. Oliphant developed a considerable refinement of techniques, particularly on the ion source. Soon after they were operating in 1933 Rutherford was sent a cubic centimetre or so of heavy water by G N Lewis, who had established a small scale concentrating plant in Berkeley. They demonstrated soon after Ernest Lawrence that the (Li6 D) reaction gave rise to a 14.5 cm group of alpha-particles, whereas the (Li7 D) reaction was found to give rise to 2 alpha-particles and a neutron with a continuous energy distribution of alphas.

More interesting still was the DD reaction leading to tritium and He-3. The interpretation of this result was illuminated by Dee's beautiful cloud chamber photographs, Dee being lent a few precious cubic centimetres of deuterium gas for use and quick return.

The next great excitement in the laboratory was produced by the Joliot-Curie discovery in January 1934, that nitrogen-13 was produced by bombarding boron with alpha-particles. We had often looked for delayed alpha-particles after bombardment of our targets but never with a Geiger counter. However, we immediately looked round the laboratory for a portable Geiger counter and were fortunate in being able to borrow one from Ken Bainbridge, who had brought one over from Harvard. There was no other portable Geiger in the Cavendish—true to its concentration on alpha-particle and proton work. We then bombarded carbon with protons for 12 minutes on 13 February 1934, and got over 70 position counts per minute—about 4 times background. The count decayed with a half-life of about 10 minutes as compared with Joliot-Curie's half-life for N-13 of 14 minutes, but it turned out that our half-life was nearer the mark. There was some disputation about our production of N-13 which some of our US friends believed were due to deuterium contamination of our beam. Before long, however, they handsomely withdrew their objections.

I have so far given a rather personal, insular account of the development of low energy nuclear physics. Perhaps the best way to make amends is to give a brief account of my tour of US nuclear physics laboratories working on these problems in the summer of 1933. I first called on Van de Graaff and was taken to see the great twin Van de Graaff generator installed at Round Hill to develop several million volts. One of the principal difficulties at that time was due to the habits of pigeons flying about in the hangar and producing from time to time high concentrations of electrical stress on the spheres. At the time we arrived the accelerating tube had not been installed. We then moved on to the Carnegie Institute to call on Tuve, Hafsted and Dahl. They had worked for a while with a Tesla coil for generating 1 MeV but abandoned this in favour of a Van de Graaff. This was at first built outside but was too much plagued with insects, so later the Van de Graaff machine was rebuilt inside a laboratory. I was particularly impressed at that time by Odd Dahl's characteristic Norwegian use of fish line for controls. The first results from the acceleration of 600 keV protons were reported at the Washington meeting of the Physical Society in May 1933, and confirmed our lithium disintegration results and also concluded that the alpha-particles emitted on bombardment of heavy elements were due to impurities.

From Washington I travelled by way of Santa Fe, the Meteor Crater and Flagstaff and the Grand Canyon to Pasadena—certainly a better way to travel than the modern six-hour trip. At CalTech I found Charlie Lauritsen with Crane working with their 1 MeV transformer and a double ended x-ray tube which looked rather similar in construction to our Cambridge tubes. Soon after my visit they turned over from accelerating electrons to deuterons and studied the production of neutrons by deuteron bombardment of lithium and beryllium.

From Pasadena I travelled to Berkeley and was enthusiastically welcomed by Ernest Lawrence and Stanley Livingston. Their 1.2 MeV cyclotron had come into operation in February 1932 and by September 1932 they had repeated our lithium work and extended the excitation curve to 750 kV. Just before my arrival they had obtained deuterium from G N Lewis and were using deuterons as projectiles and had shown that the lithium D reaction gave rise to a 13 MeV alpha-particle group as well as a continuous distribution.

They found also that an 18 cm group of protons was emitted from most of the substances bombarded by deuterons. They concluded that the deuteron was unstable in the strong nuclear field and broke up into a proton and neutron with liberation of 5 MeV of energy. This suggested a neutron mass of 1.0006—a mass which was disbelieved by the Cavendish.

As soon as I returned Walton and I started experiments with the more abundant deuterium now available from my ten dollar purchase. We found a 14 cm proton group from most elements which was evidently Lawrence's 18 cm group, but having been fooled by contamination ourselves in our first experiment we cleaned our target and showed that the group was certainly due to a film of carbon, probably due to grease. So after some long range arguments the hypothesis of the deuteron break up was abandoned.

I was particularly impressed in Berkeley with the method of operating the cyclotron. The experimenters were divided into shifts: maintenance shifts and experimenters. When a leak or fault developed in the cyclotron the maintenance crew rushed forward to plug the leaks by melting the numerous wax joints and fixed the fault when the operating shifts rushed in again.

A letter from Oliphant asked me whilst in Berkeley to purchase or steal a gallon or so of heavy water, since it was needed desperately for dozens of experiments. He added kindly that the Prof. [Rutherford] would willingly spend some money on it. So I collected about a gallon of 2 per cent D_2O and paid G N Lewis $10 for it. After passing through the Customs with difficulty, since they did not understand why I should be importing a liquid which looked very like water, I presented the heavy water and the bill personally to Rutherford, only to find that he considered the price to be too high and I should have asked his authority for purchasing it. I then tried to persuade Rutherford to allow us to build a cyclotron but owing to the success of his 200 kV accelerator he believed at that time that it was unnecessary to go to higher energies. So we had to wait till three years later to get the go ahead. By this time the acquisition of £250 000 by Rutherford for the laboratory as an appreciation for a peerage bestowed on Lord Austen by Mr Stanley Baldwin had loosened the purse strings in the Cavendish and we were also able to go ahead with the construction of a new High Voltage Laboratory. After a very considerable argument with Oliphant as to whether to build a 2 MeV home-made oil immersed cascade generator, those of us who from previous experience disliked oil immersed apparatus, which had frequently to be dismantled, won the day and we installed two Philips cascade generators and accelerating tubes. The High Voltage Laboratory under Dee's leadership produced a very interesting series of results on proton experiments, particularly resonance capture of protons. It also proved to be very useful in the war years when left in the hands of Feather and Bretscher to carry out work for the atomic energy project and it continued to produce good results until its dismantling in 1960.

2.3   A Personal View of the Cavendish 1923–30

Norman de Bruyne

My first experience of the Cavendish Laboratory was in 1923 when as a freshman I went there for lectures and practical work in physics. The building was then about fifty years old so it looked much as a 1931 laboratory might look to a present day freshman—i.e. out of fashion but not entirely obsolete. In fact, as I subsequently discovered, it was still very much Maxwell's creation as described in *Nature* on 25 July 1874 with additions made in 1896 and 1906.[1] Some of the windows had a wide flat ledge on the outside so that the sun's rays could be reflected into the room by a heliostat; the interior was partly panelled in varnished pitch pine. The outside had gothic ornamentation reminiscent of a Victorian rectory. Over the entrance gateway was an inscription 'Magna opera domini exquisita in omnes voluntates eius' which I found difficult to translate (though all Cambridge entrants had at that time passed an exam in Latin) and the King James bible only added to its incomprehensibility: 'The works of the Lord are great; sought out of all those that have pleasure therein' (*Ps* 111:3). A better translation might be 'The works of the Lord are great; they are studied by all who delight in them' which is similar to the wording in the proposed American prayer book.[2] Incidentally, D A Winstanley, a syndic of the Cambridge University Press and Vice-Master of Trinity College, once remarked to me that he thought that Lord Rayleigh (who succeeded Maxwell on his death at the age of 48) showed a surprising lack of humour in choosing this inscription for his own collected works! Whenever I entered the Cavendish I was awed by a strong sense of tradition and I would not have been surprised if on turning a corner I ran into Maxwell himself.

Practical physics was taught to freshmen by G F C Searle who as a boy had known Maxwell and who said that he had been very kind to him, but I never dared to question him about the nature of the kindness. He also knew Oliver Heaviside, a Victorian scientist martyred by the Royal Society, who according to Searle brewed his tea every week to avoid a daily chore; now that we have microwave ovens Heaviside's technique may become popular. Searle always stressed that the second law of thermodynamics did not necessarily apply to living creatures. He was an ardent antivivisectionist. In his class he adopted an air of ferocity and was particularly hard on women students one of whom was ordered to remove her stays when having trouble with a tangent galvanometer—a traditional device based on the erroneous belief that the magnetic field in a laboratory is constant. Searle always referred to Rutherford as Professor Radium who was an endearing character in a children's comic weekly called *Puck*. Searle looked exactly like the professor in Conan Doyle's *The Lost World*. Although 'cruelly precise in his denunciations' in his practical class, after I had been in Cambridge for some years he would jump off his cycle when he saw me and ask 'You-all-right?' (spoken as one word). The experiments in his class were usually home-made, unique, and sometimes irritating as when one was required to measure the surface tension of water by applying a pressure difference computed from the temperature of hot air. Each experiment was fully documented in a folder which described in detail how the experiment was to be done. Professor Norman Feather, a fellow freshman at Trinity College, said many years later about Searle's textbooks:[3] 'In their own peculiar idiom they are minor classics, the testament of a great teacher wholly committed to the task of initiating his students into the stern discipline of precise measurement in physics'. We came as schoolboys and we left as budding professionals, ready to do experiments in our third year in the advanced practical class in which we had to find our way without detailed instructions. There were two demonstrators to assist Searle but the most important person after Searle was a charming young man named Tilly who knew every experiment inside out. Whenever there was any technical trouble Searle would bellow 'Tilly!' Clarence G Tilly eventually became chief laboratory assistant and retired in about 1966. The lab assistants (men and boys) were a body of devoted helpers; Millikan (younger son of R S Millikan) told me that what he would miss most on his return to the USA would be the services of the lab boys.

Sir Joseph Thomson ('JJ') who was formerly Cavendish Professor of Physics and later Master of Trinity College occasionally gave lectures: he had a rather disconcerting habit, like President Carter, of relaxing into a smile when he paused in his speech and it made one think that one had missed a joke. However I have been told that this was merely his way of making a necessary adjustment to his ill-fitting false teeth. JJ was of course world famous for his discovery of the electron which he proved to be a particle; later on his son Sir George Thomson proved that it also had wave properties, much to the delight of his father. JJ was a man whom I was fortunate enough to see frequently, after I became a Fellow of Trinity. He had from the back a rather Chaplinesque appearance, accentuated by a walking stick which he waved behind him while he gazed intently into a shop window. His taste in shop windows was catholic and there was much speculation as to whether he was looking at the contents or at infinity. I believe he was looking at the contents because he once remarked that one of the best social changes in his lifetime had been the growth of multiple shops where working girls could get cheap pretty dresses—a much better observation than that of an Archbishop of Canterbury who thought that the most beneficial invention in his life had been that of the self-wicking candle. He came to the opening of my company in 1934 when he was 78 and said 'I hope, de Bruyne, that you will make a lot of money because Charles Parsons tells me that to do so is the mark of a good engineer'. So poor engineers remain poor!

The most entertaining lecturer was the Cavendish Professor of Physics, Sir Ernest Rutherford. His enthusiasm was infectious; sometimes it got out of control as when he once stated 'Some of the particles achieved a speed of 350 000 km/s'. This caused prolonged gentle stamping and a puzzled look on the great man's face. His weekly lecture was always enlivened with excellent demonstrations performed by his browbeaten assistant W H Hayles (pronounced to rhyme with Miles). He spoke with the authority and interest in his subject which arose from having largely created it. The worst lecturer was C T R Wilson, inventor of the cloud chamber. His audience rapidly diminished to one or two faithfuls. Nevertheless, to those patient and intelligent enough to interpret his speech and his handwriting on the blackboard (I was not one of them) it must have been a rewarding task as Blackett's words testify[4]—'I first met C T R Wilson just after World War I when I attended his lectures on light. His voice was not easy to follow, and his blackboard was difficult to read, but somehow I took adequate notes and those are almost the only lecture notes of my student days to which I have repeatedly returned. He had a penetrating but very simple approach to wave phenomena, in particular interference and diffraction. . .'. Bragg[5] gave a similar judgement: 'The very best material presented in the very worst delivery of any lectures I know'.

Alex Wood was a polished lecturer who among other philanthropic activities invested in good property for poor families. He taught us heat, light, sound and properties of matter.

C D Ellis lectured on electricity and magnetism, and also ran a practical class in electricity. He was a Fellow of Trinity and became my director of studies in my third year. He had come to physics in an unusual way; he was an engineer officer in the British army and was in Germany in 1914 when war was declared. He and Chadwick (who later discovered the neutron) were interned in Ruhleben. There Chadwick taught him physics. He was a friendly, kind person whom I got to know well in later years and I spent a holiday with him and his family in Austria. He once said to me that studying for the Tripos exam was like a long cross country race, and if you drop behind it is extremely hard to regain your position; it was not until I had read his obituary in *The Times* that I learned he had twice been victor ludorum at Harrow. His style of life was unconventional; on Mondays we all knew that he had spent the weekend in Paris and had only just arrived back in time for his lecture. His style of teaching was also unconventional, but helpful. Boltzmann's law of the partition of energy among gas molecules was 'Just the higher the fewer'.

It has been said that the physicists produced by the Cavendish were untrained in the use of accurate instruments. Rutherford was quite unmoved by such statements as he saw little merit in being able to read pointers on expensive equipment. At a meeting of the Cambridge Philosophical Society at which Professor Lowry announced that he had succeeded in measuring the index of refraction of some compound to one more decimal point, Rutherford rose saying that he felt sure that all present would wish to join him in congratulating Professor Lowry, and that he looked forward to the announcement next year that the professor had found it possible to take his researches yet one more decimal point further.

Another criticism sometimes made of our training was the absence of any compulsory mathematics. We were expected to be able to write down differential equations, such as that good old workhorse $a\ddot{x} + b\dot{x} + cx = 0$, but we were not expected to know how to turn the handles to solve them. We were trained to be horny-handed experimental physicists and those who wanted to be theoretical physicists could take the Mathematics Tripos. Not everyone was happy about this but it suited me as I have always, with some reservations, approved of Edison's arrogant remark 'When I need a mathematician I hire one'. Sir James Chadwick once paid me the doubtful compliment of singling me out in a discussion at Liverpool University as proof that physicists do not have to be good mathematicians.

One course of lectures specially interested me; it was on thermionics but alas I cannot recall the name of the lecturer. It had always puzzled me that Alexander Fleming (a Cavendish man) had done everything possible with the thermionic valve (or tube) except putting a grid in it. But the lecturer explained that any scientist would reject the idea of a grid because the current flowing to it would be in proportion to the number of lines of force ending on the grid and there would be no amplification. Only an uneducated person like de Forest would insert a grid. In fact, however, the electrons are so swift that they do not follow the lines of force; the moral to be drawn is that Edisonian research (try everything on the shelf) has its merits.

Having failed to get a 'First' in Part One of the Natural Sciences Tripos I resolved to spend two years on Part Two and this had an interesting consequence. In the first of the two years the practical class was run by Thirkill (later Master of Clare College) and in the second by Blackett who brought a breath of fresh air into the class by substituting new material for some obsolete projects. However, as Ratcliffe later remarked to me, it was not all gain; for example one of the axed projects was to construct and use a capillary electrometer from nothing more than a piece of glass tube (soft soda glass, not borosilicate) and some platinum wire. It was an excellent introduction to the realities of research at the Cavendish Laboratory. Blackett showed us how to make a Fabry–Perot interferometer from plate glass (not of course float glass) and ball bearings as separators and how to perform Millikan's determination of the charge on an electron using a Wimshurst machine as the source of high voltage; the Van de Graaff generator had not yet been invented. A quartz capillary tube bent into a U filled with mercury and momentarily heated at the bottom of the U-tube made a splendid mercury arc. In those days no one worried about mercury poisoning, the Cavendish must in some places have been saturated with mercury. Nor for that matter did anyone worry much about radiation, though the hands of Crowe (who made up all the radioactive preparations) and of Dear (a laboratory assistant) were a sorry sight.

Most of my contemporaries had spent the Long Vacation in 1927 in the 'Nursery' (the attic) of the Cavendish laboratory learning the art and craft of radioactivity under the doleful eyes of Chadwick. Professor S Devons, after describing the ordeal of the 'Nursery' on page 29 of the reference in note 1, continues: 'Later in the year I (together with a fellow student, G J Neary) was given a problem by Ellis to examine the radiations (beta and gamma) from radium C'. The experience gained in the Nursery had not been wholly inappropriate. As a research novitiate, either one had to make one's own apparatus, using hand (or foot) operated tools and bits of metal and wood that had been used and reused by generations of research students, or one might inherit and make do with the residue of some earlier research. And of course one was expected to do one's own glass blowing'. This was the Cavendish sink-or-swim system founded on the concept that physics was a gentleman's occupation that did not require any highly skilled manual abilities. I decided to spend the Long Vacation at the research laboratories of the (British) General Electric Company (GEC) at Wembley where I had spent the summer immediately before coming up to Cambridge as a freshman. I am happy to have this opportunity to record *in piam memoriam* my gratitude to C C Paterson (later Sir Clifford Paterson) and his 'Boswell' Bernard P Dudding for giving me such a wonderful opportunity to learn high vacuum techniques under Norman Campbell and B S Gosling. Professor R W Ditchburn[6] writes 'About 1926–7 mercury diffusion pumps began to appear. I think that the Cavendish was rather too slow in using these pumps, which gave a large improvement both in speed and in final vacuum obtainable. Rutherford and Chadwick were doing experiments which did not require high vacua and probably Rutherford was not keen on their expense, which for that time was far from negligible. There was, however, a general failure in the laboratory to realise what the diffusion pump could do and how to use it!' As Professor Ditchburn indicates the Cavendish was desperately poor; everything had to be done on a shoestring or with string and sealing wax. The laboratory possessed one cathode ray tube which was used by J A Ratcliffe and one day it burned out; I well remember Ratcliffe's anxiety as he stood outside Rutherford's room as to what the outcome would be.

This brings to mind a character who, to a young researcher, was more important than Rutherford himself. His name was Fred Lincoln the chief lab assistant who had the keys to all the cupboards and who had been a lab boy when JJ was the Cavendish Professor, when money was even harder to come by than in Rutherford's reign. He had a moustache with waxed twisted ends which made him resemble a recruiting sergeant of the Edwardian era. Many are the tales told of his cheese-paring harassment of research students. I remember one innocent newly arrived from some far corner of the Empire asking Lincoln for sealing wax. While Lincoln, as a gesture of good-will, was looking for a few short lengths and odd bits, the supplicant magnified his felony by adding 'Bank of England wax please'; consternation spread across Lincoln's face as he rejoined 'Bank of England! We don't live on that plane here'. I once went down to the workshop to ask Fred to order a quantity of open-ended unsealed lamp bulbs from GEC and I told him they cost about a penny each. He said nothing but at that moment Rutherford came in and Lincoln asked him 'Sir, am I to get five dozen lamp bulbs for de Bruyne?'. Rutherford bellowed, for all to hear, that it was time that de Bruyne realised the cost of his research. So I ordered them myself and the cost was zero thanks to the generosity of the General Electric Company. I later told Lincoln that we had got to face the fact that we should be seeing quite a lot of each other and that it might be less painful if we treated one another with mutual courtesy; this so delighted the man in charge of the workshop (alas I have forgotten his name) that he invited me to tea at his home for the next Sunday. He was the only man permitted to use the sole power driven lathe, all the other lathes were driven by treadle power! He told me how pleased he was that some one had at last stood up to Lincoln.

I returned to Cambridge at the end of September 1927 and set up my equipment from Wembley in a tiny room on the ground floor, with a window too high up to be able to look out on Free School Lane. It was at the remote end of the 1896 extension to the Cavendish and abutted the Department of Colloid Science run by Professor Eric Rideal. To get to my room one had to go through a much bigger room used by Allibone and by Walton and occasionally by Cockcroft who was working with Kapitza. I had decided to try for a Prize Fellowship at Trinity College by examining the effect of high electric fields on thermionic and field emission. Rutherford was my supervisor for the PhD degree and he used to visit me on Saturday mornings. Though he could not give me much help with my technical problems his conversation was always good value. Professor Alexander said 'It is curious that two of the greatest men of our days were both boys. Einstein was a merry boy until sobered by recent tragedies and Rutherford was a rowdy boy'. Rutherford might begin by asking me 'Do you Patagonians still dye yourselves with woad?' but sometimes his boisterous sense of humour verged on bad taste as when he asked 'It's a pity JJ does not wash more isn't it?' About Einstein he remarked 'He is one of those chaps that Eddington has been pushing as a world wonder'. His attitude to theoretical physicists was usually one of slightly amused disrespect, hence his ruling 'Don't let me catch anyone talking about the Universe in my department'. But in public debate with his friend Eddington, in Maxwell's lecture room, Rutherford was the loser as Eddington deftly avoided the boisterous bull charging down on him.

Eddington gave a course of lectures on relativity which made everything crystal clear to me at the time but invariably I found myself floundering some weeks later. When asked 'Is it true that you and one other man are the only people in England who understand relativity?' he hesitated rather a long time and the questioner urged him not to be modest. Eddington then replied 'No it's not that; I was trying to think who the other man was'.

If all one knew of Rutherford was that he was the talkative man in the next seat in an aeroplane or train you might easily conclude that he was a self-made businessman who had done very well. But those who worked with him were astonished by his genius. Ellis[8] wrote 'It is more than thirty years since I worked in the Cavendish and saw or talked to Rutherford almost daily, but I still feel that extraordinary dominance that he exerted over the experimental work in which he was interested. . . . Rutherford had true and simple greatness and this lay in him and not in his achievements. He seems to me to have been essentially an artist who happened to use laboratory apparatus instead of paints and canvas. On the positive side I learned from Rutherford, on his Saturday visits, the importance of generosity about other people's work; not to go round as a mini angel-of-light, making suggestions to others who had either already thought of and rejected them or were just about to think of them and, above all, to avoid the kind of biting criticism which caused Lord Brougham to devour Thomas Young.

During my time at the Cavendish I got to know Blackett better; he was approachable and friendly and treated me as an equal. We had a common interest in technology especially aircraft. I took him one day to the GEC laboratories at Wembley and he invited me to his home in Bateman Street. He was amused that I was born in Punta Arenas, known to him, when a naval officer, as 'The Paris of the South'. His ship took part, on 8–9 December 1914, in the battle, off the Falkland Islands, in which the Scharnhorst, Gneisenau, Leipzig and Nuremberg were sunk. But there was a political barrier between us. I could not attune myself to what seemed to me to be an oversimplified left-wing outlook and when I argued he always replied 'We can't put the clock back', which as Hayek has said 'Expresses the fatalistic belief that we cannot learn from our mistakes, the most abject admission that we are incapable of using our intelligence'.

I believe I first met Wynn-Williams when I had got tangled up in an electronic circuit; he brought to bear on it an obvious mastery and power of clear thinking. I next heard that he had not only repaired the automatic exchange which Cockcroft, after much persuasion, had got Rutherford to install, but had gone on to connect it, with PO approval, to the Post Office system. He then invented the ring counter followed by the scale-of-two counter, but none of us, and perhaps not even Wynn-Williams himself, realised that he had made the greatest contribution to computation since mankind found it possible to count with fingers and toes. The concluding words of his Duddell Medal address[9] to the Physical Society in 1957, reprinted in chapter 3.5, are characteristic of the man.

One Saturday morning I showed Rutherford my results establishing that the field emission was independent of temperature—contrary to Millikan's findings. I asked Rutherford if I should publish my results and he replied 'Yes if you are sure of your results but don't be surprised if you get a rocket back'. But no rocket arrived. My hazy recollection is that Millikan visited the lab some time later and when I asked him whether he was still working on field emission he replied that it was a worked out subject. However I got a rocket from Rutherford in the form of an order to cooperate with H C Webster in an investigation of a claim by an American professor that alpha-particles interacted with an electron space charge, Rutherford said it seemed inconceivable but someone would have to repeat the experiment.

I gave assistance, but to spend my time repeating dubious work was not at all to my taste so I concluded that the time had come for me to bid farewell to Cavendish. My attitude was much like Blackett's: 'If physics laboratories have to be run dictatorially. . . I would rather be my own dictator'.[10] Also the final report of my school science teacher, who had put up with me for three years, strengthened my resolve: 'Of his ultimate success I have no doubt though I think it will be in applied rather than pure science'.

Notes

1   *A Hundred Years of Cambridge Physics* (2nd edition 1980) (Cambridge: Cambridge University Physics Society, c/o The Cavendish Laboratory, Madingly Road, Cambridge)
2   The wording in the proposed American Prayer Book is: 'The deeds of the Lord are great; they are studied by all who delight in them'. I am indebted to Mr P H Summers for this information
3   Feather N 1975 Twice Thirty Years of Physics *Contemp. Phys.* **16** 489
4   1975 *Biog. Mem. Fell. R. Soc.* **21** 6
5   Allibone T E 1966 Reminiscences of Fifty Years Research *Proc. R. Soc.* A **41** 93
6   Ditchburn R W 1967 Vacua at the Cavendish *Phys. Bull.* **28** 566

7 Allibone T E 1964 The Industrial Development of Nuclear Power *Proc. R. Soc.* A **282** 448

8 Ellis Sir Charles 1960 Rutherford – One Aspect of a Complex Character *Trinity Review* (Lent Term)

9 1957 34th Duddell Medal Address *Phys. Soc. Year Book* 53

10 1975 *Biog. Mem. Fell. R. Soc.* **21** 22

2.4   Reminiscences 1930–4

W E Duncanson

As one of Chadwick's research students from 1930–4 I can probably add a few reminiscences to those of the more distinguished people at the Cavendish at that time. In order to illustrate the atmosphere of the place and the spirit that pervaded the Cavendish during that period, I shall give a few glimpses of life in the laboratory.

As soon as one joined the Cavendish one was conscious of being accepted into a friendly community. This meant a great deal to those of us who came from overseas. This friendliness was not limited to one's fellow research students but was shown also by the senior people, including those who were not directly concerned with one's work. For example Wynn-Williams and Webster were particularly helpful to me, the former giving me much advice and help on the devising of equipment needed to detect protons in the presence of an intense gamma-ray background. Webster handed over to me some valuable equipment he himself had constructed—in those days research students had to make most of their own equipment. These are two personal examples; more generally I should like to mention the many important contributions John Cockcroft made to the work of the Cavendish, often in an unobstrusive manner. Many can testify to this and also to his interest in and consideration for other people. A personal experience is but a small illustration of this. In our experimental work we were using a radioactive source (RaB+C+C') which has a half-life of the order of about 20 minutes and, therefore, we had to make the most of the short time available with each source we prepared. When Cockcroft, working in an adjoining room, realised that the spasmodic sparking over of his high voltage equipment played havoc with our experiment he would switch off until we had completed our observations. This was no mean concession as he and Walton were on the verge of carrying out the first disintegration experiments with artificially accelerated particles.

These personal examples may appear trivial but they illustrate the attitude that permeated the laboratory. Perhaps a more outstanding case is that of Rafi Chaudhri who arrived in Cambridge in September 1930 at about the same time as myself but in a more unorthodox manner. It was commonly reported in the Cavendish that one Sunday afternoon he turned up at Rutherford's home, complete with luggage, informing Rutherford that he had come to work with him. It can be understood that Rutherford was not too pleased. It appears that Chadwick and Oliphant persuaded Rutherford not to turn him away and Oliphant suggested that he would take Chaudhri as one of his research students. Oliphant continued his support when Chaudhri returned to India and, later, was instrumental in gaining equipment grants for Chaudhri on his appointment to a professorship in Pakistan.

Other members of the Cavendish were helpful to Chaudhri in a number of personal ways. I recall especially Philip Moon and Jack Constable who befriended him and gave him considerable assistance when problems arose such as the difficulties he had with landladies when they were required to provide breakfast at 4 AM due to a midsummer Ramadan!

Earlier I mentioned the inspiration Rutherford gave to those who worked under him. P I Dee, quoted by Crowther, summed this up as follows: ' Rutherford would always be remembered by those who knew him for his boisterous and friendly personality and for his happy and total absorption in his own work and that of his collaborators.'

Rutherford kept in touch with all research in the laboratory; he would periodically walk round chatting to people about their work, revealing a detailed knowledge of what was being done even by those not under his direct supervision. He also held strongly to the opinion that research students should not be slaves to their work but should take advantage of the wider opportunities Cambridge offered. For this reason the Cavendish closed at 6 PM. I recall an occasion during the long vacation when I was still working after most others had gone off on holiday. After discussing my work he suggested it was time I had a break; it was not an instruction but I took the hint.

Rutherford's friendly nature did not prevent an occasional furious outburst. During my first year I was closely associated with Jack Constable, who was working with Chadwick. They were using a parallel-plate ionisation chamber which unfortunately acted also as a sensitive microphone resulting in spurious effects if anyone spoke loudly. Constable's patience had its limits, leading to his putting up, in a prominent place, an illuminated notice saying 'Shut up, damn you' which was switched on at appropriate times. Unfortunately, one occasion was inappropriate as Rutherford, speaking in his booming voice, had just entered the room on his way to his own laboratory. His reaction was predictable resulting in a deflated Constable. On another occasion dark clouds hung over the laboratory for about a week: Rutherford was accustomed to having his senior people successful when they applied for professorships and had taken it for granted that Blackett would be offered the Chair of Natural Philosophy at Aberdeen. He was infuriated when

Carroll, one of Eddington's men, was appointed to the post. However, these were passing clouds and, apart from a temporary cautiousness in the laboratory, they did not upset the general happy atmosphere of the place.

Before I move from stories of Rutherford, one more may not be out of place. From time to time the Rutherfords held lunch parties to which a number of research students would be invited; these often coincided with the visit of some distinguished guest from abroad. I was present at one such luncheon. At one end of a long table sat Lady Rutherford while, at the other end, sat Rutherford and his guest, with the research students distributed along the two sides. An incident that occurred exemplified Dee's remark about Rutherford's total absorption in his work. Once he became involved in a discussion everything else took second place; and so it was on this occasion. Rutherford, with scarcely a break in the animated discussion with his guest, passed up his plate and said 'more pudding please'. As there was very little steamed pudding left, Lady Rutherford was not very pleased and, with some annoyance, said 'Ern, you will wait'; she then asked the guest whether he would like some more, 'No, thank you'; 'Dr Feather would you like some more?' 'No, thank you'; 'Mr Cairns?', 'No, thank you'; and so on round the table with the same reply; finally she said, 'Ern, pass your plate' and proceeded to divide the small remaining piece of pudding into two halves, sending the remaining portion back to the kitchen.

I shall now turn to matters associated with the discovery of the neutron. As an introduction I shall mention a light-hearted meeting held by younger members of the laboratory some time in 1931. It was a caricature of the Cambridge Philosophical Society and was billed as a meeting of the Phoolosophical Society. I have forgotten all but one of the papers read at the meeting; this particular one was given by Ted Nicoll, a Canadian research student, and was entitled 'The Fewtron'. The main thing I remember about this paper was that the fewtron seemed to have few, if any, properties—in fact, its only property seemed to be a negative one, namely, that if no tracks appeared on a photographic plate after exposure to a Wilson cloud chamber expansion, then you knew that a fewtron had passed through the chamber!

As I have already indicated there were, at the Cavendish, people considerably senior to me who were more directly concerned with Chadwick's work on the neutron but I can provide some personal information on earlier work which, by chance, had a bearing on Chadwick's neutron experiment. Riezler, a research student from Germany, had made measurements on the scattering of alpha-particles by some of the light elements. At about the same time there became available the results of Chadwick, Pollard and Constable on the disintegration of aluminium by polonium alpha-particles; the protons emitted indicated two alpha-particle resonance levels. Riezler, therefore, designed and partially constructed equipment with a view to detecting these levels by large angle scattering of alpha-particles by aluminium. Riezler had to return to Germany so Chadwick suggested that, while I was awaiting the delivery of a piece of equipment, I should join him in completing this experiment.

The relevance of this work to Chadwick's later neutron experiment was that a polonium alpha-particle source was required, the preparation of which Chadwick always did himself. Under the conditions prevailing at the time the preparation of such a source was not a simple matter and the ultimate strength of the source was somewhat unpredictable. It happened that, on this occasion, the source was the strongest Chadwick had ever prepared. Although this strong source was of considerable help to us, we soon realised that the geometry of our equipment was not sufficiently precise to enable us to resolve the more recently discovered four alpha-particle resonances. Chadwick, therefore, suggested that I should return to my initial project with Harold Miller, who had recently joined me. At about that time Curie and Joliot had obtained some puzzling results when they examined 'radiation' emitted from beryllium when exposed to alpha-particles from a polonium source. Chadwick said to me that he would like to examine further this work of Curie and Joliot.

It was a happy chance that Chadwick's polonium source was so strong, as it made a considerable difference to the certainty of his results and to the time taken to complete the conclusive demonstration of the existence of a new particle, the neutron. It was all done in the compass of two weeks.

I shall conclude these recollections with a few personal memories of Chadwick as I knew him. As already noted, Dee described Rutherford as boisterous and totally absorbed in his work. One could not, by any means, describe Chadwick as boisterous but he did have the same dedication to his work. I still have vivid memories of the period when we were working on the scattering of alpha-particles; he broke the Cavendish rule of 6 o'clock closing by continuing up to 10 PM and sometimes later. I particularly remember one occasion when, at about 7.30 PM he said to me 'I must go home as we have some people coming to dinner, but I shall be back later'. Some time between 9.30 and 10 PM he returned. I do not remember what time we finished nor did I hear what his wife had to say!

The more one worked with Chadwick the more one appreciated his encouragement, friendliness and concern for those working under him. I know that many people found him difficult and some even described him as rude, but this reaction by some people arose from his intense enthusiasm for his work. Often when I went into his room to ask a question or to discuss some matter he would completely ignore me, continuing at his little table doing a calculation or writing some notes. After five, ten or even fifteen minutes I would depart and there would still be no recognition that I had been in the room. On other occasions he would note my presence, in which case I would say my piece but get no response; again I would make myself scarce. Later—maybe several hours later—he would come out of his room and discuss the point I had raised without any recognition of a lapse of time.

One quickly adjusted to these idiosyncracies and I found Chadwick a very helpful and challenging personality. In his own way he tried to bring the best out of people and to develop their confidence. I experienced this personally in many ways but particularly in the research problem he allotted me and in his request to me, on one occasion, to lead the Cavendish colloquium. Chadwick was in charge of the fortnightly colloquium at which some recently published paper would be discussed. I was given the formidable task of introducing Heisenberg's initial paper on nuclear structure in which he postulated the recently discovered neutron as one of the nuclear constituents. Not only was this paper highly theoretical but it was also in German— I had little qualification on either score to do the job but somehow I managed to get through it. Fortunately, when questions came from the many theoretical physicists in the audience Dirac, much to my relief, very considerately took control of the discussion.

One could instance many other occasions when Chadwick gave encouragement to his research students but enough has been said to indicate the contribution he made to the reputation of the Cavendish. Those of us who worked with him owed him a debt of gratitude for his leadership and inspiration.

2.5   Research at the Cavendish 1932

Harrie Massey

At the beginning of the 1931–2 session the Cavendish Laboratory was working well with plenty of enthusiasm but no expectation that a period of
exciting discovery to match the heady years of the birth of wave mechanics
was soon to come.

Much of the experimental research was directed towards the study of the
disintegration of nuclei by alpha-particle impact. No other effective projectiles were available. Resonance levels had been discovered and there was an
awareness that the disintegration of beryllium exhibited some features which
had not been properly understood. Improved techniques, involving electronic methods of counting, were coming into operation in the laboratory
and Chadwick was taking advantage of these to elucidate the beryllium
problem. Rutherford himself was concerned with alpha-particle energy levels
in nuclei, a subject which impinged somewhat on Ellis's investigations of
the wavelengths of the gamma-rays emitted in beta-decay. More than four
years had passed since Ellis and Wooster had carried out their remarkable
experiment on the average energy of disintegration of radium E and,
although this revealed the problem of the energy relations in beta-decay in
stark clarity, there was remarkably little talk about this problem in the
Cavendish at the beginning of 1932. Perhaps this was because of the complete
puzzle which it presented. Instead Ellis was concentrating most of his attention on attempting to make quantitative measurements of internal conversion
coefficients of gamma-rays.

Blackett was already beginning to work with Occhialini on the design and
construction of a cloud chamber in a magnetic field which would be triggered
by a cosmic event. Other cloud chamber experiments were in progress in
the laboratory, some concerned with studies of the scattering of fast electrons
by nuclei. On some of the photographs tracks of electron–positron pairs
undoubtedly appeared but were disregarded, the concept of the positron
being viewed as a theoretical gimmick. This is not surprising for it really
was an 'out of this world' concept at the time.

A further set of experiments which proved to be highly significant for pair
formation were those of Gray and Tarrant on the absorption of gamma-rays

by heavy elements. They were finding anomalously strong absorption for gamma-rays of quantum energy much in excess of 500 keV. That is to say the absorption was greater than would be expected from the known processes—the photoelectric and Compton effects.

Perhaps least noticed of all concerned with the study of atomic nuclei were the experiments of Cockcroft and Walton. They were attempting to generate DC voltages to accelerate protons to sufficient energy to prove useful as projectiles in nuclear disintegration experiments. It was felt at the beginning of 1932 that they had a long way to go.

In addition to the experiments in nuclear physics, if we may use so anachronistic a term, there were also a number concerned with atomic physics. After all it was only six years since Schrödinger's classic papers on wave mechanics had appeared. Most of these experiments were housed in a large ground floor room known as the 'Garage'. It was in this room that, with E C Childs, I began the extension of the work which Bullard and I had carried out on the diffraction of slow electrons by gas atoms, to scattering by metal vapours. Most of the other experiments in the Garage were concerned with positive ions. R M Chaudhri was working on the interaction of positive ions with surfaces, B M Crowther on the electromagnetic separation of isotopes and E T S Shire on an experiment much in advance of its time. Following his successful cloud chamber experiments on the scattering of alpha-particles in helium which revealed the effects of indistinguishability of the colliding nuclei predicted by Mott, Blackett suggested that Shire should attempt to observe these effects for scattering at keV energies. As an indication of the difficulty of such a programme it may be noted that the effects were not observed at these energies until 1965, by Lorents and Aberth.

The back of the Garage was reserved for J J Thomson who was carrying out experiments looking for what would now be called plasma oscillations. He came in for a few hours just about lunchtime each day.

For those of us whose experiments required the establishment of a high vacuum (at the time 10^{-6} mm Hg was a high vacuum) 1932 was the first year in which we were able to use Pyrex instead of soda glass for our vacuum envelopes and associated equipment. The relief was very great even though it was quite hard to track down an oxygen cylinder for Pyrex glass blowing. Even though relative modernity was creeping in to the nuclear counting work the laboratory budget was so small that almost all basic equipment was in short supply.

Mohr and Nicoll were beginning their remarkable experiments on the angular distribution of slow electrons scattered after suffering a definite energy loss in a gas, which were not repeated until more than 30 years had elapsed. They had an experience which exemplified both the good and bad features of Pyrex. Their elaborate experimental chamber in Pyrex glass with a Pyrex ground joint had just arrived from Felix Niedergezass, the laboratory glass blower. While making an adjustment to the metal electrode system enclosed, a screwdriver slipped and smashed a clean triangular hole in the Pyrex cylinder. They came down to the laboratory tearoom (a small room opening into the library) looking woebegone. However, Childs and I had had plenty of experience joining Pyrex, as in our apparatus, which had to operate at a high temperature, the electrode system was carried on a short closed cylinder. After adjustment this was fused to the main cylinder using an oxygen flame. We volunteered to join up the piece of Pyrex which had been knocked out and somewhat to our surprise this was achieved sufficiently well to preserve the vacuum properties of the envelope and enable the experiments to proceed. With soda glass this could not have been done even by good glassblowers which we were most certainly not.

Although it was not one of his main interests Rutherford paid regular visits to the Garage and would sit down smoking his pipe and talk about what progress was being made, as well as many other things. Chadwick, as Assistant Director of Research, took a general interest in everybody's work. He was very understanding about the requirements of the 'high vacuum' experimenters. Rutherford was apt to regard the use of Hyvac pumps for long periods during baking of the vacuum equipment as very wasteful of laboratory resources but Chadwick would always calm him down.

A further line of work was the investigation, by P Kapitza, of the effects of strong magnetic fields on the properties of materials. This again enjoyed Rutherford's strong support even though it was outside his main interest in physics and very far from string and sealing wax in practice.

Research on the ionosphere was also being carried out, under the direction of J A Ratcliffe who had already established himself as a world authority on the subject.

Thus even in 1932 the laboratory was carrying out research on what would now be called nuclear physics, atomic physics, solid state physics and atmospheric geophysics. It was not difficult, however, to follow with interest the progress of colleagues working on the different aspects. It was easy to wander into any laboratory—there were no locked doors and some rooms, like the Garage, housed several experiments.

Every second week there was a meeting of the Cavendish Physical Society on Wednesday afternoons. Tea on such occasions was distinguished by the fact that it was donated by the Rutherfords and Lady Rutherford poured out the tea personally. At these meetings a member of the laboratory would give an account of some new work for which he was responsible. Rutherford always presided and had some remarks to make which added to what was always a very interesting occasion. It is hard to convey the air of excitement at the prospect of new revelations in physics.

Apart from this there existed three clubs, the $\nabla^2 V$, the Kapitza and the Junior Physical. Membership of the first two, which met in college rooms, was limited and decided by election. The third included all research workers in the laboratory, mainly those in their first or second year of research, who were not members of the other two clubs. All three meetings, which were quite informal, consisted of the presentation of a paper either by a member of the laboratory, or by a visiting physicist, followed by usually extensive discussion.

I was fortunate in that, through carrying out theoretical as well as experimental research, I was a member of the theoretical research group which worked rather loosely under the supervision of R H Fowler, Rutherford's son-in-law. The existence of this group marked an important break away from the traditional schools of applied mathematics in being directly associated with an active physics laboratory. Although I was confined to atomic physics for experimental research there was no limitation on the theoretical subjects which could be tackled. I was already much involved with the quantum theory of collision processes. Much of this was concerned with atomic collisions but I was also working on the anomalous scattering of alpha-particles. It was easy then to keep in touch with the nuclear experimental work.

Nevertheless the discovery of the neutron by Chadwick in February 1932 was kept very quiet for some little time until it was all very thoroughly confirmed. A full account was first given by Chadwick at a meeting of the Kapitza or $\nabla^2 V$ Club. To this meeting he invited his old friend J G Crowther who was one of those rare birds at the time, a science correspondent (for the *Manchester Guardian*). For Crowther it was somewhat of a scoop and there was widespread interest by the Press generally.

In the 1920s Rutherford had speculated on the possible existence of atoms of zero nuclear charge arising from very tight binding of an electron to a proton, and their role in the building up of heavy nuclei. In the period immediately after the discovery of the neutron all that was known of its mass was that it was nearly equal to that of a proton. Rutherford's picture, which would have required that the neutron should have a smaller mass than the proton, was not at first ruled out (although it was later on). Even though according to non-relativistic quantum mechanics the lowest state of an electron in the field of a charge $+e$ is the 1S state, at the time one could not be sure that a fully relativistic two-body theory would not yield a much lower state. In default of any other basis I applied atomic collision theory to discuss the passage of neutrons through matter on the assumption that it was essentially a very strongly bound H atom, but in view of later more accurate measurements of the neutron mass this was largely a waste of time. However, in the course of this work I had many interesting talks with Chadwick which made the work seem worthwhile.

Hardly had the excitement over Chadwick's discovery died down than another unexpected major discovery occurred. It was also kept quite confidential for some time and I first heard of it in the Cavendish Laboratory after tea one day. Mott came in and said to me 'Have you heard that Cockcroft and Walton have done this.' Forbearing to say it in words he wrote down

$$Li + H \rightarrow 2He.$$

Few of us had expected any physics from the work of Cockcroft and Walton until they had doubled the working voltage to 500 kV or more. Yet here a whole new world of nuclear physics experiments had been opened up, beginning with an accelerating voltage as low as 250 kV.

Euphoria over this discovery developed as it appeared that much heavier elements could also be transmuted by impact with 250 keV protons. Judging from the experiments there seemed no limits to this but it became increasingly difficult to believe. Mott was able to obtain theoretical limits for the total cross section for collision between systems with prescribed relative velocity and angular momentum. Fowler asked me to investigate the chance that a heavy element could react with protons to the apparently observable extent in terms of a simplified model including resonance levels. Fortunately I was able to avoid a tedious, unrewarding calculation because, using Mott's result, limits could be attached to the contribution from a resonance level. These limits showed that the observed results from heavy elements were far greater than the theoretical. It was soon found that the boron in the Pyrex glass used in the experimental arrangements was responsible because of its high reaction cross section. The laboratory then settled in to a systematic investigation of transmutations by proton impact. Rutherford persuaded Oliphant to apply his experimental skill to this subject, leading to the later experiments with deuterium.

While all this was going on Blackett and Occhialini were making steady progress in the development and operation of their counter controlled cloud chamber. Not only did they make this technique work but they observed positron–electron pairs, formed among the secondary products of cosmic rays, which they were able to identify with very little ambiguity.

There has never been a year in which so much of fundamental importance was discovered in one laboratory. All who worked at the Cavendish at the time were aware of this and felt most fortunate to have been so close to it all, even if not directly involved.

Part 3

Underlying Themes

3.1 Introduction

Three features of the background to the events of 1932 stand out as being of special interest, and as suggesting fairly radical revisions of what might be called the myth of the Cavendish. Under the direction of Rutherford the laboratory is often thought of as having been unconcerned with, and even hostile to, recent theoretical developments in physics. It is also often conceived of as having been biased against industry and industrial sponsorship, and conservative in its attitude towards advanced apparatus and instrumentation.

There is undoubtedly a good deal of truth in this traditional picture. The keynote of the laboratory, and of Rutherford's work in particular, was simplicity, and this ideal had a profound influence upon attitudes to both theory and experiment. Rutherford himself was often impatient of complex mathematical theory, preferring to think rather in terms of simple physical images and encouraging his colleagues to do likewise. He had also made his own remarkable discoveries with only the crudest of apparatus, and again expected others to follow his example, using their brains rather than their wallets. The general policy on apparatus has been described as 'extremely conservationist'. Expenditure was not encouraged, and researchers were expected to improvise their own apparatus from a limited store of available materials. The Cavendish researchers were indeed, as Lord Bowden has recalled, 'the most impecunious physicists I ever knew'.[1] So far as industry was concerned Rutherford's attitude was again simple, and again cautious. Money received from industry threatened the freedom of research, while help given to industry was also discouraged, as in Rutherford's advice to Kapitza: 'You cannot serve God and Mammon at the same time'.[2]

The myth, then, is not without foundation, but as the three famous experiments themselves illustrate, there is also another side to the story. Though generally cautious of involvement with industry, Rutherford did in fact develop very close links indeed with one industrial firm, and without these links neither the work of Cockcroft and Walton, nor the large programme of research on magnetic fields carried out by Kapitza, would have been possible. So far as new instrumentation was concerned a small group working under Chadwick's direction in the late 1920s and early 1930s developed a set of instruments that put the Cavendish for a short period far ahead, technically, of any of its rival laboratories. Both Chadwick's own discovery of the neutron and Blackett's investigation of the positron rested heavily upon recently developed instrumentation and techniques.

The suggestion that the Cavendish was hostile to theory is also misleading. The Cambridge organisation did not make allowance for a department of theoretical physics such as those which had thrived for some years in Copenhagen and Göttingen and nurtured the creation and development of the new quantum mechanics. Theoretical physics came under the Faculty of Mathematics, and the connections between the theoreticians and the Cavendish experimentalists were for the most part informal. But connections there were, and again both the Cockcroft–Walton and Blackett experiments were closely related to, and even dependent on, recent theoretical developments.

Theory and experiment

The idea that the Cavendish was hostile to theoretical physics may be traced to three major factors. One of these was of course Rutherford's emphasis on simplicity, leading to an impatience with complex mathematical theory. A second was the frequent absence abroad of the leading Cambridge theoreticians, Fowler and Dirac, making supervision of theoretical research students something of a problem. The third was the absence of any formal relationship between the theoreticians of the Mathematics Faculty and the experimentalists of the Cavendish Laboratory. Reflecting the force of two of these circumstances Alan Wilson, who was a research student in the Mathematics Faculty, has recalled his frustration at the limited opportunities for supervision and collaboration:[3]

> The contact between theoretical physics and experimental physics was minimal, the former being in the Faculty of Mathematics and the latter in the Faculty of Physics and Chemistry. The only point of contact was through R H Fowler—because he was Rutherford's son-in-law. But Fowler, like the rest of us, worked in his college rooms—in Trinity—and if you wanted to consult him you had to drop in half a dozen times before you could find him in. He lived in Trumpington and did most of his work there. Also he spent half his time in America.
>
> Dirac was unapproachable and also spent a lot of time abroad.
>
> There was nowhere in the Cavendish where one could meet people and exchange views. There was, in fact, a seminar once a week on Wednesday afternoons lasting an hour for research students to present as well as they could some of the enormous number of papers appearing in the *Zeitschrift für Physik* and the *Annalen der Physik*, but with Dirac and Fowler often absent the discussion of these was perfunctory.

Wilson's reflections, more of which are included in his contribution below, are not bitter. Nor are they the complaints of a student in whom no one was interested. He was, as we shall see, taken into the select circle of the Kapitza Club, and his research on the theory of semiconductors resulted in a classic contribution to that field. But he was able to make significant progress only after spending time on the Continent, where the supervision and facilities for discussion were far better than in Cambridge.[4]

By 1932, as Wilson indicates below, things had improved considerably. Following the transfer of mathematics teaching from the colleges to the University in 1926 the mathematicians had become gradually less isolated, and the upsurge of interest in theoretical nuclear physics in the wake of quantum mechanics had brought them much closer to the physicists of the Cavendish. However, there were still no Mathematics Faculty buildings, and no regular colloquia in theoretical physics. Both Fowler and Dirac were still abroad for much of the time (including a good part of 1932) and their absence, Dirac's renowned remoteness, and the absence of facilities for theoreticians in the Cavendish remained very real problems for the theoretical physics research student in Cambridge. Moreover, the third of the mathematical professors to be in reality a physicist, namely Eddington, was also a remote figure, and while friendly with Rutherford and Fowler at Trinity College, and with Stratton and C T R Wilson at the observatories, he seems to have had very little contact with the Cavendish itself. Once this point has been made, however, it must also be admitted that Wilson's experiences of the late 1920s were unfortunate and not altogether typical of the early 1930s. There were contacts between theory and experiment, even for students, and among the leading physicists of the period, the already established researchers, such contacts were often very close.

The most formal of these contacts, and so far as research students were concerned the most important, was through Ralph Fowler, whose presence suffered from his foreign travels but was nevertheless strongly felt. Originally a pure mathematician, Fowler had been enormously influenced by Rutherford's arrival in Cambridge. He and Rutherford had become fast friends immediately, and he had then married Rutherford's daughter, turned his attention to theoretical physics, and taken an office alongside Rutherford's in the Cavendish. His own direct contributions to physics were, as noted in Part 2 above, considerable; but they were probably outweighed in importance by his contribution as a teacher and as a link with the tremendous developments in theoretical physics that had taken place on the Continent in the 1920s. It was Fowler who in 1923 communicated Louis de Broglie's famous paper on the wave theory of matter to the *Philosophical Magazine* and Niels Bohr's definitive survey of the old quantum atomic theory and its problems to the *Cambridge Philosophical Society*.[5] Dirac studied under him, and learnt through him of Heisenberg's profound 'new kinematics' of 1925, the starting point of his own major contributions to quantum mechanics.[6] In the early 1930s, having himself made significant contributions to the field of astrophysics opened up by his Cambridge colleague Eddington, and hav-

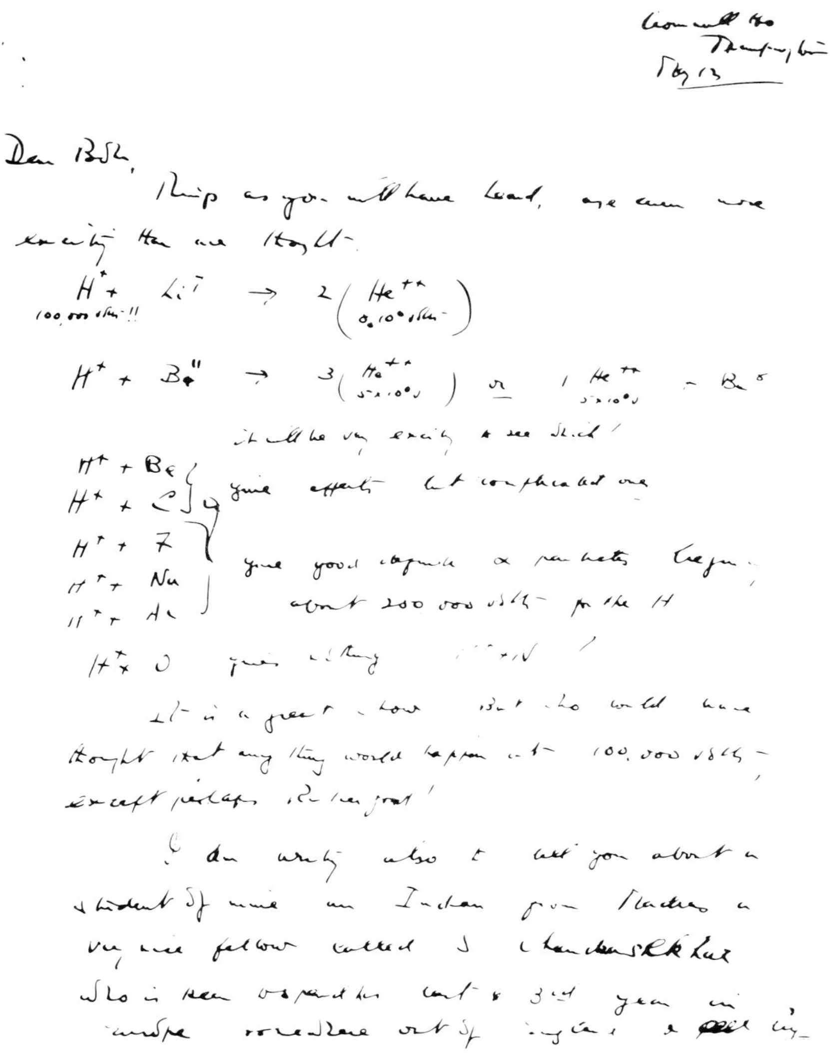
Dear Bohr,

$H^+ + Li^7 \rightarrow 2(He^{++})$

100,000 volts!!

It is a great show. But who would have thought that anything would happen at 100,000 volts — except perhaps Rutherford!

I am writing also to tell you about a student of mine an Indian from Madras a very nice fellow called S Chandrasekhar

ing in the process encouraged his then colleague E A Milne, Fowler acted as sponsor to a generation of young astrophysicists, including Chandrasekhar. In this period too he directed a group of theoretical physics research students that included both Nevill Mott and Harrie Massey. Significantly, both Mott and Massey played an active part in discussions on the experimental work in the Cavendish, while Massey conducted experimental as well as theoretical research. Through this work they became fully integrated with the laboratory, although they were nevertheless primarily theoreticians.

Figure 3.1.1 (Opposite) Letter from Fowler to Bohr from the Bohr correspondence in *SHQP*, courtesy AIP Niels Bohr Library. A transcript of the letter is given below.

Dear Bohr,

Things as you will have heard are even more exciting than we thought.

$H^+ + Li^7 \rightarrow 2 \begin{pmatrix} He^{++} \\ 8 \times 10^6 \text{ volts} \end{pmatrix}$

100,000 volts!!

$H^+ + B^{11} \rightarrow 3 \begin{pmatrix} He^{++} \\ 5 \times 10^6 v \end{pmatrix}$ or $\begin{pmatrix} 1\ He^{55} + Be^8 \\ 5 \times 10^6 v \end{pmatrix}$

it will be very exciting to see which

$H^5 + Be$ } give effects but complicated ones
$H^+ + C$ }

$H^5 + F$)
$H^5 + Na$ } give good definite α particles beginning
$H^5 + Al$)   about 200,000 volts for the H

$H^+ + O$ gives nothing $H^+ + N$?

It is a great show! But who would have
thought that anything would happen at 100,000 volts,
except perhaps Rutherford!

I am writing also to tell you about a
student of mine an Indian from Madras a
very nice fellow called S Chandrasekhar . . ..

So far as many of the staff of the Cavendish were concerned, some of the most important contacts with theoretical physics were probably the quite informal ones operating through college relationships and personal friendships. Blackett's work on the positron has been described by the philosopher N R Hanson as 'the perfect marriage of experiment and theory',[7] and Dirac's brief recollections in Part 1 of this volume confirm that the theoretical side of this marriage was not merely an appendix to Blackett's experimental investigation, but rather integral with it. Whereas Anderson discovered the positron almost by chance, Blackett, working closely with Dirac, knew from the beginning just what he was looking for, and was so able to apply stringent tests to determine whether or not he had found it. The image of the Cambridge experimenter is often based on a picture of Rutherford, working alone and applying the simplest of concepts to the simplest of situations and experimental arrangements. By the 1930s, however, it was perhaps Blackett who provided the archetype of the experimentalist, as described by himself in an essay of 1933 on the craft of experimental physics:[8]

The experimental physicist is a Jack-of-all-trades, a versatile but amateur craftsman. He must blow glass and turn metal, though he could not earn his living as a glass-blower nor even be classed as a skilled mechanic; he must carpenter, photograph, wire electric circuits and be a master of gadgets of all kinds; he may find invaluable a training as an engineer and can profit always by utilising his gifts as a mathematician. In such activities will he be engaged for three-quarters of his working day. During the rest he must be a physicist, that is, he must cultivate an intimacy with the behaviour of the physical world. But in none of these activities, taken alone, need he be pre-eminent, certainly not as a craftsman, for he will seldom achieve more than an amateur's skill; and not even in his knowledge of his own special field of physics need he, or indeed perhaps can he, surpass the knowledge of some theoretician. For a theoretical physicist has no long laboratory hours to keep him from study, and he must in general be accredited with at least an equal physical intuition and certainly a greater mathematical skill. The experimental physicist must be enough of a theorist to know what experiments are worth doing and enough of a craftsman to be able to do them. He is only pre-eminent in being able to do both.

Blackett was truly a Jack-of-all-trades, no theoretician himself but with enough command of theory to collaborate effectively with the leading theoreticians. He was not alone in this. James Chadwick was also a competent theoretician, for example, and so was John Cockcroft. Cockcroft actually did very little original physics, for his real talent lay in organisation; but his range of attributes, like Blackett's, was broad. Although he was a prominent member of the Cavendish Laboratory, physics was in fact the one subject in which he was not qualified. His first degree had been in engineering, after which he had worked as a graduate apprentice at the Metropolian-Vickers Company. He had then read mathematics at Cambridge, spending much of his time working in the Cavendish Laboratory but nevertheless emerging as a senior wrangler (i.e. with a First Class degree). It was through this familiarity with mathematics that he was equipped to follow the development of theoretical physicists and so to pick up the proposals of Gamow that lay behind his famous work with Walton. It is sometimes suggested that there was at the Cavendish a certain lack of faith in the new quantum mechanics, and this would tie in with Massey's observation, recorded in Part 2, that in late 1931 not much was expected to come of the Cockcroft–Walton work. But it must be equally significant that Cockcroft and Rutherford did have enough faith in the quantum-mechanical predictions to embark upon a relatively large programme, the success of which was dependent entirely upon the theory. Moreover, despite the strong international interest in the development of artificial sources of particle beams for nuclear research, the Cambridge project was the only one to make use of the results of theory in this way.

Apart from sparking off the accelerator programme, the young Russian George Gamow also contributed in other ways to the theoretical awareness of the Cavendish Laboratory around the turn of the decade. It was a peculiar feature of Cambridge physics that when foreign physicists came on a visit they usually came to the Cavendish, whether or not they were experimentalists, and this naturally had a destructive effect upon the barriers existing within the Cambridge physics community. Gamow was quite definitely a theoretician, but when he first visited Cambridge it was with an introduction

from Bohr to Rutherford, and when he returned for a full academic year in
1929–30 he quite naturally spent that year at the Cavendish, working with,
among others, Chadwick.[9] Then, when he returned to Niels Bohr's Institute
for Theoretical Physics in Copenhagen, he continued to correspond with his
Cavendish friends, thus providing a link between the world's leading centre
of theoretical physics and its leading experimental physics laboratory. A
letter to Cockcroft written in March 1934 in what became known as the
English dialect of the language 'Gamowian', although not typical of their
exchanges, perhaps deserves quotation:[10]

> Dear Cockcroft!
> Thank you very much for your letter with last news from nuclear world. It is
> always a great excitation in the Institute when your letters come and a spetial
> commision of english-speaking pleople and spetialists on egiptian and babilon-
> ian discuss four hours the questions wether '*NEWson*' mean 'however' or
> 'hidrogen' and what is to be understood under the notation '*yelwson*'. After
> these difficult philological questions are setteled the tekst comes into hands of
> physicists.

In rather better English, Bohr himself also kept up regular and, in times of
rapid development, very frequent correspondences with Rutherford, Fowler
and Dirac.[11] Both Fowler and Dirac were often to be found in Copenhagen,
while Bohr was a frequent visitor to Cambridge. Through these contacts
Bohr's institute and Rutherford's laboratory became almost sister establish-
ments, and the theoretical mastery of the former provided the balance that
Cambridge on its own to some extent lacked.

Outside contacts played a significant part in keeping the Cavendish abreast
of theoretical developments, but such contacts were not restricted to the
Copenhagen group, and nor were they the only occasion for theoretical
intercourse. For within Cambridge, the main forum for the consideration of
external developments was also the main meeting point of theoreticians and
experimentalists, namely the Kapitza Club.[12] Founded by Peter Kapitza in
1922 and meeting once a week after dinner in its members' college rooms,
this quickly became one of the leading physics seminars in the world. The
original idea was to provide a forum for unfettered discussion of current
developments in physics, freed from the inhibitions that tended to charac-
terise laboratory discussions, and for theoretical debate, for which there was
no provision in the formal university system. Membership of the club was
by invitation, and was always restricted, and in the early years the members
were almost all, apart from Kapitza himself, primarily theoreticians. Thus
the prominent members during the first two years were D R Hartree,
Herbert Skinner, E G Dymond, M H A Newman, E C Stoner and J E
Jones, and the club then constituted something of an informal theoretical
physics department. As the years passed the membership increased from
single figures to between twenty and thirty, and while the club continued to

be a meeting point for the university's theoreticians a significant number of experimentalists were also admitted as members. For the 1931–2 session, for example, the membership was as follows: J D Bernal, then establishing his reputation as a crystallographer, Patrick Blackett, John Cockcroft, Philip Dee, Paul Dirac, Norman Feather, Ralph Fowler, L H Gray, Peter Kapitza, Harrie Massey, Nevill Mott, Mark Oliphant, F C Powell, John Ratcliffe, J K Roberts, C P Snow, later to find fame as a writer, T G P Tarrant, Ernest Walton, H C Webster, Alan Wilson and W A Wooster. Giuseppe Occhialini, who became a member the following session, was admitted as a visitor.

This was a prestigious seminar group, and it was one in which the barrier between theory and experiment was all but obliterated. The early meetings of the club, in the early to middle 1920s, consisted of members reporting on developments in the recent physics literature, and of discussion of these developments, and this remained a core activity. In the 1923–4 session, for example, Skinner reported on Compton's theory of x-ray collisions (the Compton effect), and later on Pauli's use of this theory to derive the statistics for the equilibrium between radiation and matter. As the club minutes record, the members were not impressed by Compton's theory, although after the discussion of Pauli's work Blackett and Stoner broke ranks and supported it. The reports of 1931 included Dymond on the work of Landau and Peierls on the relativity extension of Heisenberg's Uncertainty Principle. In May 1932 Fowler and Dirac reported back on what they had picked up the previous month in Copenhagen.

From quite early on the club also listened to reports of its members' own research, and by the 1931–2 session this was perhaps the dominant activity. Alan Wilson reported on his semiconductor studies, Walton on the progress of the accelerator, Bernal on his x-ray crystallography, Chadwick on his neutron discovery, and so on. Finally there were the talks by eminent visitors on their own research. These again started quite early in the club's history, and by the late 1920s and early 1930s they amounted to a most impressive programme. Guest speakers in the late 1920s included James Franck, Niels Bohr and Erwin Schrödinger as well as Pavlov, Nordheim, Joffé, Dennison, W Kuhn, K T Compton, Lennard-Jones, G P Thomson, C G Darwin, W H Bose and many more. The 1931–2 and 1932–3 sessions saw talks by Eddington, LeMaître, Millikan, Uhlenbeck, Bohr, Beck, Bethe, Tamm, Morse, Houtermans, Goetz, Weisskopf and, again, many more. Any suggestion that the Cavendish was isolated from current theoretical developments thus breaks down completely as soon as the Kapitza Club is taken into account.

Techniques and instrumentation

The most important development in instrumentation to take place in the Cavendish in the period before 1932 was the introduction of electrical counting methods, aspects of which are described by Norman Feather in Part 1 and by W B Lewis and C E Wynn-Williams below. As Feather notes, the first electrical particle detector had in fact been devised by Rutherford and Geiger in Manchester as early as 1908.[13] The device consisted of a metal cylinder filled with a gas at low pressure and with a metal wire, well insulated from the cylinder, passing along its axis. The cylinder was raised to a high negative potential, and the wire connected to an electrometer. An alpha-particle entering the cylinder would cause an initial ionisation of the gas, the products of which would then cause an avalanche of further ionisations. As the negative electrons which were released massed up on the central wire the event would be recorded by the electrometer. So long as the potential applied to the cylinder was in a certain range the pulse recorded by the electrometer would be proportional to the initial ionisation produced by the incident particle, and highly ionising rays such as alpha-particles could be distinguished from less ionising beta- and gamma-rays.

The new device had proved effective but not altogether reliable, and it had happened that at almost the same time the possibility of an alternative method of detecting alpha-particles, on a zinc sulphide scintillation screen, had also been raised. It had been known since the previous century that certain materials became luminescent under the bombardment of positive radiation, and in 1908 it was shown that zinc sulphide laced with 0.01% copper was not only extraordinarily sensitive in this way to alpha-particles but also luminesced in the most visible part of the spectrum.[14] As Lord Bowden relates below, the discovery of the properties of this crystal at this time was to have tremendous repercussions. From Rutherford's point of view the scintillation screen was every bit as effective as the electrical detector, more reliable, and much simpler. Moreover by using the screen he was able in a sense to 'see' the impinging alpha-particles, and this satisfied his strong need for a visual image of the process he was investigating. The electrical detector was accordingly abandoned.

For the next thirty years Rutherford and his team, first in Manchester and then in Cambridge, continued to work with the scintillation screen, which was of course used in the Cockcroft–Walton experiment in 1932. In Cambridge they also used the cloud chamber invented there by C T R Wilson and put to such effective use by Blackett. Like the scintillation screen the cloud chamber was also a very visual instrument, and by the early 1930s this was probably the most used tool in the laboratory. In the latter part of the 1920s, however, it had become clear that a new method of detecting and counting particles was needed. For analysing single events the cloud chamber was indispensable. But for recording and counting numbers

Figure 3.1.2 Rutherford and Bohr in Cambridge, 1930. Front row left to right: Lady Rutherford, Mrs Oliphant, Mrs Bohr. Reproduced by kind permission of Sir Mark Oliphant from his book *Rutherford: Recollections of the Cambridge Days* (Amsterdam: Elsevier, 1972).

of particles the scintillation screen had major disadvantages.[15] It was, above all, highly dependent on the observer. It did not give any objective record, a failing that was particularly significant in cases of dispute such as that noted by Feather, and since it was reckoned that for reliability an observer could be trusted only for about six hours a week it was demanding on the staff. The work of counting was laborious and tedious and the apparatus was subject to the constraint that the number and rate of particles recorded could be no higher than could be counted by the observer, perhaps one or two a second. Rutherford, with his preference for simplicity and for visual methods, does not seem to have strongly encouraged the development of electrical alternatives but he did recognise the need for this development and was content to see it carried on in the Cavendish under Chadwick's supervision.

The first significant development in this field in fact came from Germany, where Hans Geiger had maintained an interest in devices such as that he had constructed with Rutherford many years earlier. In 1928, working with W Müller, Geiger developed a new electrical detector which immediately became known as the Geiger–Müller counter, or, more briefly, the Geiger counter.[16] This was basically similar to the detector of 1908, but whereas the ionisation-dependent response of the earlier device had made it optimal for alpha-ray detection the new apparatus was more suitable for the detection of the less ionising beta- and gamma-rays. In crude terms, the potential was raised above that of the original detector, with the result that the proportionality between the initial ionisation and the resulting discharge broke down, and all ionising particles entering the cylinder gave rise to similar discharges.

The Geiger–Müller counter was very sensitive, being able to react even to a single ionisation anywhere in the tube, but it also had its drawbacks. The way in which the initial ionisation avalanched into a full discharge was not fully understood, and this introduced an element of uncertainty into the use of the device. Until artificial means of satisfactorily quenching the discharge were developed in the mid-1930s (one such being devised in the Cavendish by Wynn-Williams) there was also a strong tendency for a discharge to be followed by further, spurious, discharges, unrelated to any incident particles. Thus although Chadwick responded to the new invention by quickly building several himself for use in the Cavendish the extent of such use, mainly by H C Webster, was limited.[17] The detectors were more reliable, however, when used for coincidence counting. Then the effect of the spurious discharges was greatly cut down, as the chance of two such discharges exactly coinciding was small. The use of Geiger counters in this way was of course crucial to the experiments of Blackett and Occhialini.

The main Cavendish effort in the field of electrical detectors was spearheaded by C E Wynn-Williams, F A B Ward, and later W B Lewis, and was directed towards the construction of large-gain linear amplifiers with which to detect alpha-particles and protons as ionisation pulses in a shallow ionisation chamber. The first steps in this direction had been made on the Continent by Greinacher, Ramelet and Ortner and Setter,[18] but the idea was quickly taken up and much developed and improved in the Cavendish. In 1929 Ward, Wynn-Williams and H M Cave used a small ionisation chamber linked to a linear valve amplifier and thus to an Einthoven galvanometer to record the rate of alpha-particle emissions from radium.[19] The following year Chadwick, J E R Constable and E C Pollard used similar apparatus to study the relationship between the energies of incident alpha-rays and emitted protons in nuclear transmutations, and this apparatus was again used by Chadwick two years later in his discovery of the neutron.[20] In each case the particles to be detected entered a very shallow ionisation chamber, whose metal casing was connected to the positive pole of a 1000 volt battery, and whose inner electrode, on which the positive ions produced by the impinging particles collected, was connected to the grid of the first valve in the amplifier circuit. For the device to be effective this valve had to have a very low inter-electrode capacity and very good grid insulation, but these properties were found to be available, as Lord Bowden recalls below, in the Marconi Company's DEV valves. The amplifier was completed by a circuit containing four further valves, and this was linked to the Einthoven string galvanometer which recorded the pulses produced on a cylindrical chart. Since the amplification was linear the deflection on the chart was proportional to the ionising power of the impinging particle, allowing this to be measured and different types of particle to be distinguished.

Meanwhile, in the course of another sequence of experiments, Ward and Wynn-Williams had in 1930 devised a major modification to the apparatus incorporating improved circuitry and a double ionisation chamber with oppositely charged electrodes set one behind the other.[21] The advantage of this arrangement was that particles of selected ranges could be isolated from a background of other particles, even if this background were very strong. The energy spectra of particle emissions could therefore be studied. Up to a certain range, the particles would produce a net ionisation in one direction, but thereafter the effect of the second, larger, chamber would outweigh that of the first and the ionisation would have the opposite sign. The amplifier reacted only to ionisation of one sign. Other improvements to the chamber and the circuitry brought the response time of the apparatus down to the order of a thousandth of a second, and this meant that the effects of nearly coincident particles could be isolated sufficiently to distinguish accurately between single strongly ionising particles and the superposition of the effects of several more weakly ionising particles, even when the background of such weakly ionising particles was very considerable. Already in 1930, Rutherford, Ward and Wynn-Williams, analysing the spectra of short-range alpha-particles from naturally radioactive elements, could isolate separate groups of rays amounting to as little as 1 in 4000 of the total radiation

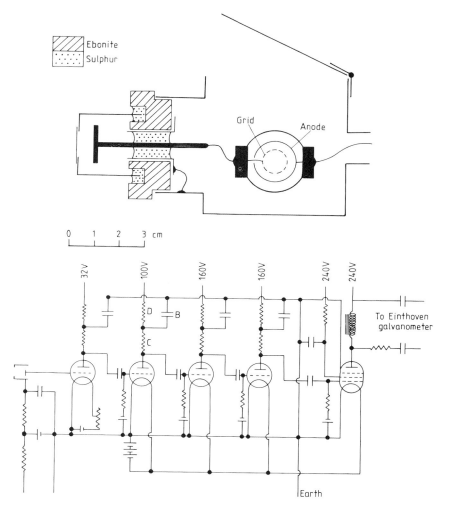

Figure 3.1.3 The detection apparatus developed by Ward and Wynn-Williams and used by Chadwick in his discovery of the neutron. *Above*: ionisation chamber. *Below*: linear amplifier. Redrawn from *Proc. R. Soc.* A **125** 717 (1927) and **130** 465 (1930).

entering the chamber, and could count separately as many as a hundred or more events per second.[22]

One final development of counting instruments in the period culminating in 1932 was not in fact significant for the famous experiments of that year, but was itself of considerable historical importance and illustrates well the pioneering atmosphere in the Cavendish in that period. It concerned the development by Wynn-Williams of the scale-of-two counter. The key component of the new development was the thyratron, a gas-filled triode which

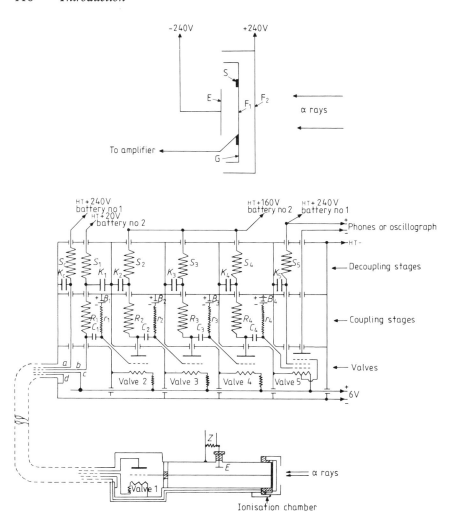

Figure 3.1.4 Double ionisation chamber and improved amplifier devised by Ward and Wynn-Williams in 1930. *Above*: double chamber counter. *Below*: diagram of amplifier and ionisation chamber. Redrawn from *Proc. R. Soc.* A **129** 217 (1930) and **131** 403 (1931).

acted as an ordinary vacuum triode until the applied voltage reached a certain value at which point it arced between the anode and the filament. The effect was that of an inertialess relay, and in 1929 A W Hull of the General Electric Company in America suggested that it might be used in a counting mechanism.[23] Following a visit by Hull, H C Webster and Norman de Bruyne in the Cavendish took up this idea and used the thyratron as a relay for counting

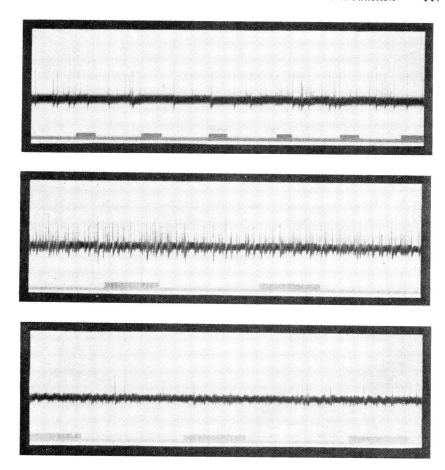

Figure 3.1.5 Charts produced by linear amplifier detectors. The time marks are at one-second intervals. *Top*: 8.6 cm (short kicks) and 4.8 cm (long kicks) ranges of thorium using single ionisation chamber. *Middle*: differential chamber—positive chamber alone. *Bottom*: differential chamber—positive and negative chambers. From *Proc. R. Soc.* A **129** 217 (facing) (1930).

the pulses from a Geiger counter.[24] The device worked satisfactorily, but the speed of counting was limited by the speed of the mechanical counter and was inadequate for the very high rates for which visual counting was either impossible or laborious, and for which an automatic counting mechanism was really necessary.

The first step towards improving the thyratron circuit was taken by Wynn-Williams in 1930 and the improvement was put to use by Rutherford, Wynn-Williams and W B Lewis during the following year.[25] The idea, described below by Wynn-Williams, was to put together a number of thyr-

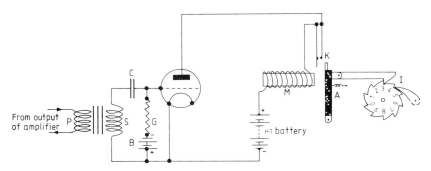

Figure 3.1.6. Single thyratron counting circuit. Redrawn from *Proc. R. Soc.* A **132** 298 (1931)

atrons in a ring. When the same voltage was applied to all of them only one could respond by arcing, and a circuit was devised whereby this response affected the grid current bias of the next in the ring, ensuring that this would respond to the next impulse and so on. The introduction of further relays or condensers to the circuit enabled the thyratrons to be automatically reset after arcing, so that the process could be carried on indefinitely. The counter was then linked up to one of the thyratrons so that with, for example, a ring of five thyratrons the counter would be triggered by every fifth impulse. In this way the mechanical limitation of the counter was overcome. The overall counting rate could be improved, and the interval within which consecutive pulses could be recorded, being dependent on the electrical properties of the circuit rather than on the counter, could be dramatically reduced. In the 1931 apparatus impulses received within 1/500 of a second of each other could be effectively recorded, even though the mechanical counter took 1/25 of a second to operate.

The thyratron ring constituted a major advance in counting technology, but it was still very limited. The total rate of particles that could be recorded was limited to the capacity of the counter multiplied by the number of thyratrons. More significantly, the apparatus was complex, and not easily reproducible. The cathodes of the thyratrons had to be heated, and separate power supplies and accumulators were needed. Later in the year, Wynn-Williams therefore proposed a new circuit which had none of these major drawbacks.[26] Instead of using a ring of thyratrons he put together a cascade of units, each containing a pair of thyratrons. The idea was that each thyratron of the initial circuit would respond to alternate impulses. One of these thyratrons was then linked to the second circuit, each thyratron of which would itself respond to alternate transmitted impulses, or to one in four of the original impulses. After a number of stages one of the thyratrons in the final unit, recording, say, every eighth or sixteenth event, was linked up to a counter. The great advantage of the new system, apart from its potentially higher capacity, was that the two-thyratron circuit was much simpler than the thyratron ring. The whole apparatus could be run off the laboratory mains electricity supply, and the need for heating and separate accumulators was also obviated. Wynn-Williams's first scale-of-two counter could record impulses received within as little as 1/1250 of a second, and this was subsequently improved by modifications made by W B Lewis in 1934.[27]

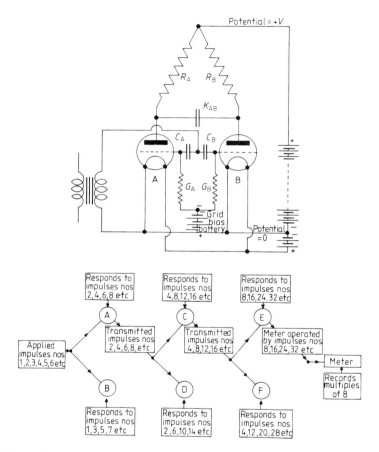

Figure 3.1.7 Single two-thyratron circuit and cascade arrangement for three two-thyratron units. Redrawn from *Proc. R. Soc.* A **136** 314–5 (1932).

The full potential of the scale-of-two counter was not realised until after a new design based on ordinary valves rather than on the expensive and troublesome thyratrons had been developed by W B Lewis in 1937,[28] and, indeed, until it came to be used in the early development of computers immediately following the Second World War. But Wynn-Williams's invention, which was in operation by early '1932, still stands as one of the great developments of instrumentation in physics, and as yet another great achievement of that remarkable year in the Cavendish.

Academe and industry

If Rutherford's attitude to developments in instrumentation was cautious, his attitude to involvement with industry, and especially to industrial sponsorship, was far more so. Very conscious of the importance of freedom in academic work he had no wish to place himself under any obligation, real or imagined, to an industrial benefactor. Unwilling to have to justify his application of money and resources he would rather go without them, no real hardship when he believed anyway that invention flourished best when starved of material aids.[29] Circumstances did not always support this stand, however, and by the mid-1920s a number of connections had already been forged betwen the Cavendish and the laboratories of the Metropolitan-Vickers Company at Trafford Park in Manchester. Some of these were in the form of straightforward commercial relationships, as the work of Kapitza in particular required expenditure on large scale electrical machinery. Other relationships were less formal, and led to an intricate interweaving of the work of the university and the firm.[30]

The main link was through John Cockcroft, who had been a graduate apprentice at Trafford Park before going to Cambridge, and who remained on the firm's pay-roll as a consultant. This consultancy was, moreover, far more than a merely nominal one, for in visits back to Manchester and in voluminous correspondence with George McKerrow, who was responsible for general liaison between the firm and the universities, Cockcroft provided much valuable advice and information.[31] McKerrow in turn reported on any developments within the firm's laboratories that might be useful or of interest to the Cavendish.[32] Sharing a room with Cockcroft and Walton in the late 1920s, Thomas Allibone was also a Metropolitan-Vickers man, seconded from the firm to study under Rutherford for a doctorate, but remaining entirely supported by the firm rather than by the university. When Allibone returned to Trafford Park in 1930 his place in the Cavendish was taken by yet another of the Metropolitan-Vickers engineers, Brian Goodlet. Other members of the Cavendish, including Chadwick, Ellis, and G I Taylor, who spent the summer of 1932 sailing round the Outer Hebrides with McKerrow,[33] also had contacts with the company.

As Allibone recalls below, the importance of these contacts for the work of the Cavendish was considerable. The Cockcroft–Walton accelerator project, generally thought of as simply a Cavendish exercise, in fact owed much to Metropolitan-Vickers assistance. In retrospect it may seem rather a quaint indication of the simplicity of approach of the Cavendish that the vacuum seals in the apparatus were made from Plasticine. But at the time Cockcroft and Walton set up their apparatus Plasticine, which quickly came to replace sealing wax as the standard sealant, was very much a new development, and one which made a considerable difference to high vacuum techniques. The Plasticine, known for trademark reasons as Apiezon sealing compound Q, was supplied by Metropolitan-Vickers where it had been developed along with a whole range of Apiezon oils and greases by C R Burch, and in the summer of 1931 there was still some doubt as to whether it would be available in the quantity required (about 6–8 pounds) by Cockcroft and Walton.[34] The pumps for the apparatus were also produced by Burch, while the transformer was designed by Goodlet, again at Metropolitan-Vickers, especially for the experiment. After the successful disintegration of lithium Rutherford decided that he must have his own accelerator with which to explore the possibilities opened up, and the transformer for this, used by Rutherford in his work with Oliphant, was again supplied—though this time on a commercial basis—by Metropolitan-Vickers.[35]

Although Rutherford himself had connections with the Metropolitan-Vickers Company, arising from contact with their director of research, A P M Fleming, in Manchester during the First World War, and although he would visit Trafford Park when pressed to do so by Cockcroft and Allibone, he seems to have been rather cooler about the relationship than were his staff.[36] He apparently remained cautious too about accepting money, as a letter from McKerrow to Cockcroft written in February 1932 suggests:[37]

> After careful consultation with the 'parties infected at T. Park' it has been decided that the best way to square this is by sending a cheque for the amount to the Cavendish for 'services rendered'. Please see the Crocodile [Rutherford] doesn't send it *back*!

When it came to the work of Peter Kapitza, who was the author of the nickname 'Crocodile' (it was based on Russian folklore), the sums involved in dealings with the firm were very much higher than the few pounds concerned here. Whereas the research of the rest of the Cavendish was conducted on a shoestring, Kapitza's experiments on high magnetic fields were both big and expensive. By the 1920s they were being funded by the DSIR, but the contract for the large short-circuit alternator that formed their centre-piece was put out to Metropolitan-Vickers. The relationship through the DSIR was of course a formal one, but as well as conducting his own research John Cockcroft also spent a lot of time working with Kapitza, generally keeping his laboratory in some sort of order, and he provided a direct link between Kapitza and the firm and ensured that everything went smoothly—at least in so far as anything in which Kapitza was involved could go smoothly. When Kapitza convinced Rutherford to persuade the Royal

Figure 3.1.8 Kapitza with his heavy-current machine. Reproduced by kind permission of Sir Mark Oliphant from his book *Rutherford: Recollections of the Cambridge Days* (Amsterdam: Elsevier, 1972).

Society to allocate a very large sum of money from the Mond bequest for a new cryogenic laboratory in which he could prepare pure crystals at very low temperatures, Metropolitan-Vickers were again contracted to supply much of the apparatus. At about the same time, around 1933–4, the firm also supplied a large electromagnet for Blackett, enabling him to generate a continuous high magnetic field for use with his automatic cloud chamber. Many other smaller items were also supplied to various members of the laboratory.

Although these latter examples were cases of commercial transactions, the bond that had been built up between the Cavendish and Metropolitan-Vickers ensured that they were handled rather differently from the norm. As Allibone's recollections below indicate, the relationship between the two institutions was a quite uncommonly close one, and one that was absolutely central to the success of the Cavendish in the early 1930s.

Notes

1   Lord Bowden 1979 Lecture delivered at the University of Canterbury, New Zealand; see also Oliphant M L 1972 *Rutherford: Recollections of the Cambridge Days* (Amsterdam: Elsevier)

2   Kapitza P 1967 *Collected Papers* volume 3 (Oxford: Pergamon) p. 237
3   Letter to the Editor 9 October 1981
4   Wilson A H 1980 *Proc. R. Soc.* A **371** 39
5   de Broglie L 1924 *Phil. Mag.* **47** 446; Bohr N 1923 *Suppl. Proc. Camb. Phil. Soc.*
6   Dirac P A M in C Weiner (ed) 1977 *History of Twentieth Century Physics* (New York: Academic). For Fowler's career see *Obit. Not. Fell. R. Soc.* (1948) **5** 14
7   Hanson N R 1963 *The Concept of the Positron* (Cambridge: Cambridge University Press) p. 163
8   Blackett P M S in H Wright (ed) 1933 *University Studies* (London: Nicholson and Watson) p. 67
9   See Gamow G 1970 *My World Line* (New York: Viking Press)
10   Gamow to Cockcroft 29 March 1934 Cockcroft Collection, Churchill College, Cambridge
11   Most of this correspondence is in *SHQP*
12   Minutes of the Kapitza Club, Cockcroft Collection, Churchill College, Cambridge and *SHQP*
13   Rutherford E and Geiger H 1908 *Proc. R. Soc.* A **81** 141
14   See the papers of Feather in Part 1 above and Bowden below.
15   See for example Chadwick J, Constable J E R and Pollard E C 1931 *Proc. R. Soc.* A **130** 463
16   Geiger H and Müller W 1928 *Phys. Z.* **29** 839
17   See Sargent B W *Recollections of the Cavendish Laboratory Directed by Rutherford* (Department of Physics, Queen's University, Kingston, Ontario)
18   Greinacher H 1926 *Z. Phys.* **36** 364; Greinacher H 1927 *Z. Phys.* **44** 319; Ortner G and Setter G 1929 *Z. Phys.* **54** 449; Ramelet E 1928 *Ann. Phys.* **87** 871
19   Ward F A B, Wynn-Williams C E and Cave H M 1929 *Proc. R. Soc.* A **125** 713
20   Chadwick J, Constable J E R and Pollard E C 1931 *Proc. R. Soc.* A **130** 463
21   Wynn-Williams C E and Ward F A B 1931 *Proc. R. Soc.* A **131** 391
22   Rutherford E, Ward F A B and Wynn-Williams C E 1930 *Proc. R. Soc.* A **129** 211
23   Hull A W 1929 *Gen. Electr. Rev.* **32** 397
24   de Bruyne N and Webster H C 1931 *Proc. Camb. Phil. Soc.* **27** 113
25   Wynn-Williams C E 1931 *Proc. R. Soc.* A **132** 295; Lord Rutherford, Wynn-Williams C E and Lewis W B 1933 *Proc. R. Soc.* A **133** 351
26   Wynn-Williams C E 1932 *Proc. R. Soc.* A **136** 312
27   Lewis W B 1934 *Proc. Camb. Phil. Soc.* **30** 843
28   Lewis W B 1937 *Proc. Camb. Phil. Soc.* **33** 549
29   See for example the Bowden and Oliphant references mentioned in note 1 and the Kapitza *Papers* (note 2) pages 22 and 237

30 See Niblett C A 1980 Images of Progress *PhD Thesis* (University of Manchester)
31 Correspondence between McKerrow and Cockcroft, Cockcroft Collection, Churchill College, Cambridge
32 Correspondence between McKerrow and Cockcroft, as note 31
33 McKerrow to Cockcroft, 19 July 1932, Cockcroft Collection, Churchill College, Cambridge and Niblett (note 30)
34 Bowes to Cockcroft, 1 June 1931 and McKerrow to Cockcroft, 27 February 1932, Cockcroft Collection, Churchill College, Cambridge
35 Allibone to Cockcroft, 23 June 1932, Cockcroft Collection, Churchill College, Cambridge and see Allibone's paper below
36 See Niblett reference (note 30)
37 McKerrow to Cockcroft, 3 February 1932, Cockcroft Collection, Churchill College, Cambridge

3.2   Theory and Experiment at the Cavendish circa 1932

Nevill Mott

I entered St John's College, Cambridge, to read mathematics in 1924. As both my parents had worked in the Cavendish Laboratory under J J Thomson, I had little doubt that my ultimate objective was physics, but with a scholarship in mathematics, it seemed that I should aim to do theoretical rather than experimental research. I felt however that I should have some experience of laboratory work, so I went to a long vacation course in Dr Searle's laboratory. G F C Searle had known my parents, but I don't think he had much opinion of my capacity for practical work, and I didn't stay the course. Stories about Searle were many; the one I like the best was of a student from a far country, stronger perhaps in applied maths than in experimental physics, who asked him for an infinite solenoid and got the reply: 'You'll have to go to the National Physical Laboratory for one of those; we don't have them here.'

In those days most mathematical scholars omitted attending lectures for Part I of the Tripos, starting with those for Part II, taking three years over them and also, as an option, going to advanced lectures up to the research level in the third year. This advanced work was called 'Schedule B', and success in it added a B or a B-star to one's class in Part II. This has now been replaced by Part III. Together with another student, John Brunyate of Trinity, I decided to take Part II and the Schedule B examination at the end of my second year, to free my third year for the beginnings of research. We were both successful in getting a B-star, and I spent some of the long vacation with a family in Germany, because I knew I would need the language to read the papers on the new quantum mechanics.

I remember meeting Brunyate at the beginning of the next term. Schrödinger's equation had recently appeared, and Brunyate's opinion was that it had taken all the fun out of quantum mechanics, using as it did the technique of differential equations, which was 'old hat'—rather than matrixes, non-commutative algebra and the methods of the original papers of Heisenberg and Born and others. Though still undergraduates, we were free to do research, and we went off to see R H Fowler, to see if he could suggest a problem. At that time theory had no recognised place in the Cavendish. No space was assigned to theorists, who were supposed to work in their college rooms or in their lodgings. Theorists were members of the Faculty of Mathematics. They could—and some did—sit in the Rayleigh library in the Cavendish and meet their peers there. It was also used as a tea room, and apt to be rather crumby and smelly, but friendly. The leading theorist, R H Fowler, later Sir Ralph, not yet a professor, was also in the Faculty of Mathematics, but his relationship with the department of physics was close. He was Rutherford's son-in-law, and he had a room in the Cavendish, next door but one to Rutherford's. If you wanted to see him you went up the stairs, past Rutherford's door, and (often) stood in a queue. This Brunyate and I did. But Fowler told us that he was going on leave to the United States for the winter, and we should come back in the spring. Brunyate was so discouraged by this and—as he felt—by the dullness of the maths in Schrödinger's equation, that he abandoned physics and took up the study of law, eventually becoming solicitor to the university. I had a rather solitary year, as far as physics was concerned, reading the German papers, particularly that of Max Born on collision theory, and also Dirac's papers on radiation theory. Dirac was, as far as I can remember, the only theorist of eminence who was working on quantum mechanics in Cambridge during that year; I remember once going to him for help, but on the whole I was on my own. It was, I think, a good introduction to research. I took my first research problem from Born's paper on collisions, *Wellenmechanik der Stossvorgange*. All the ideas for treating collision problems, together with the probability interpretation of the wavefunction were there, but he used an approximate method to solve the Schrödinger equation (the Born approximation). I wanted to see whether an exact solution for a particle in a Coulomb field would lead to the Rutherford scattering law and found that it did. I took it to Fowler, then back from his leave, who sent it to the *Proceedings of the Royal Society*. To sum the series of spherical harmonics multiplied by radial wavefunctions, I used the method of steepest descents, learned from Fowler's Schedule B lecture on statistical mechanics that I had attended the previous year. A much more elegant derivation of the Rutherford law (by Gordon) appeared, fortunately for me, rather later.

Next year I was a registered research student, and after a year in Cambridge spent a term in Copenhagen, with the enormous privilege of working under Niels Bohr. Douglas Hartree and George Gamow were there at the time. Gamow was fresh from his success in explaining alpha-particle decay by quantum mechanical tunnelling. I believe it was this that convinced Rutherford that the new theories had something to be said for them. I confess that I was rather jealous; all I had done was to show that the new theory predicted Rutherford scattering, as it had to do if the theory was right. 'Ah, Motti', Gamow used to say 'you must construct an alpha-particle'.

I couldn't do that—we didn't even have a neutron to do it with—but I noticed a paper by J R Oppenheimer on the application of antisymmetrical wavefunctions to collision problems, and applied his analysis to collisions between free electrons. At 45°, the scattering should be just half that predicted by the Rutherford formula. It was Fowler who put me on to trying the analysis for alpha-particles, which obeyed Bose rather than Fermi–Dirac statistics, as was known from the band spectra of He_2; I didn't know this till Fowler told me. The scattering should be twice that predicted by the Rutherford law, if the particles were slow enough for the interaction to be Coulombic. Chadwick carried out the experiment, and when it was clearly as I'd proposed, he took me along to see Rutherford, who said, 'If you think of anything like this again, come and tell me'. That certainly made my day.

I spent the summer of that year in Göttingen, and the next academic year in Manchester with a lectureship in theoretical physics under W L Bragg. I was then offered a teaching fellowship at Gonville and Caius College, Cambridge, together with a university lectureship in the Faculty of Mathematics—and also got married—so I spent the three years 1930–3 in Cambridge, not as a member of the staff of the Cavendish but frequently talking with them. So, the theme of this essay being that of the Cavendish around 1932, I can report on my memories as a young teaching officer in mathematics who sought out subjects for research in physics.

These were the great days in Cambridge, with the discovery of the neutron and the artificial disintegration of the light nucleus, the positron and, slightly earlier on the theoretical side, Dirac's explanation of the electron spin. The problems of the nucleus and of the application of quantum mechanics to scattering problems seemed overwhelmingly the most attractive to a young theorist. Of course there were other areas of research, such as Kapitza on low temperatures and magnetic fields and Bernal in crystallography. Fowler, now a professor, still occupied his room along the corridor from Rutherford, and the students of theory, increased in number, still queued in the corridor to see him. Fowler's *magnum opus*, in cooperation with Charles Darwin, was on statistical mechanics, and his large book, published by the Cambridge University Press, showed that he had a very good knowledge of the early application of quantum mechanics to solids. None the less, as far as I know, only one of his research students worked in this field and published a few joint papers with him; this was A H Wilson (later Sir Alan), whose essay for the Adams Prize laid the foundations of our understanding of the difference between metals and insulators and of the nature of semiconductors. This came about through his interaction with the German school, particularly during a visit to Leipzig as he makes clear in his contribution to *The Beginnings of Solid State Physics*.[1] This work turned out to be of the greatest importance, as I realised when I moved to Bristol in 1933 and began myself to work on similar problems. However I don't think it made much impact in the Cavendish. I have a curiously vivid memory of Fowler explaining it to Charlie Ellis (C D Ellis) and Ellis replying 'very interesting' in a tone that implied that he was not interested at all. I myself, together with a Manchester research student, J M Jackson, published a paper somewhat related to the science of solids, on the energy exchange between inert gas atoms and a solid surface.[2] This arose from the experimental work of J K Roberts—so the work of the Cavendish, even outside the Mond Laboratory was still by no means one hundred per cent nuclear.

Among research students in theoretical physics at that time were H R Hulme, later Chief of Nuclear Research at the Atomic Weapons Research Establishment (from 1953–73), and H M Taylor, later Treasurer and then Secretary General of the Faculties in Cambridge and after that Vice-Chancellor of the University of Keele. Hulme took his PhD in 1932, with a thesis working out the lifetimes of the internal conversion coefficient of gamma-rays, using the relativistic wave equation of Dirac.[3] Taylor and I extended this to the case of a quadrupole transition. Taylor was Fowler's research student, but was assigned to me while Fowler was (again) away on leave, and we had a very fruitful collaboration. In fact, I was still a research student too, although a Fellow of a college. It seemed to me rather absurd to pay £4 a term to Fowler while I received £4 each from his students, so I withdrew my candidature for the PhD and never took the degree.

Ellis became interested in this work, and he and I published a paper,[4] communicated in December 1932, comparing these results with experiments. I remember frequent discussions with Ellis at that time, and as a result we published another paper[5] in which we examined the two routes for the disintegration of thorium C to ThPb, both of which involve an alpha as well as a beta disintegration, but in opposite orders. We were led to the conclusion that 'the energy difference between the initial and final nuclei is *equal to the upper limit of the beta-ray spectrum*'. We said we did not wish in the paper to speculate on what happens to the excess energy, but, we stated, our hypothesis was consistent with the suggestion of Pauli that the excess energy was carried off by particles of great penetrating power such as neutrons of electronic mass. These papers suggest to me now that collaboration with theorists was good in the Cavendish—to the great credit of Fowler and also a result of the experience some of us had had abroad. I remember Blackett saying to me, when I returned from Copenhagen, 'I see you've learned to work in a lab'. But it also suggests to me that young theorists, under the influence of Rutherford, Chadwick and Ellis, may have been discouraged from speculating too wildly. We really had in Ellis's work much of the evidence for the existence of a neutrino, and, with hindsight, it is perhaps a pity we didn't say so.

At that time Harrie Massey and I started working on our book *The Theory of Atomic Collisions*. Following my papers on the Rutherford scattering law and alpha–helium collisions I had done a good deal of work on collision problems—but it was Massey who really developed the subject with the second and third editions and with other books, while I turned to solid state when I went to Bristol in 1933. I confess I do not remember how we first got together; I hope Sir Harrie's contribution to this volume will remind me of it.

On Fowler's suggestion we published our book with the Oxford University Press, in the series he and Kapitza edited, and in which Dirac's *Principles of Quantum Mechanics* was the first and most famous contribution. Cambridge University Press had published in 1929 Fowler's massive *Statistical Mechanics* and some people expressed surprise that he should abandon Cambridge for Oxford with—I would guess—a much more saleable series. I remember him explaining this to me—'after all, the Oxford Press is a much bigger show'—but I was much too junior to need an explanation. Later, with Kapitza back in Russia and Fowler much taken up with other things, he suggested that I take his place as editor of the series, and the association with the OUP is one that I have very much enjoyed ever since.

By 1935, when Fowler was writing his second edition,[6] the application of quantum mechanics to solid state physics was well advanced, and the massive reviews by Sommerfeld and Bethe had appeared. That Fowler was well aware of them is apparent from his book. Perhaps his outstanding contributions were applications to our understanding of stellar atmospheres.

As regards publications of research papers, it was normal for Fowler or one of our colleagues on the experimental side to communicate all our publications to the *Proceedings of the Royal Society*, except perhaps minor ones which went to the *Proceedings of the Cambridge Philosophical Society*. I do not know why this was. Before 1925 Fowler's more important papers appeared in the *Philosophical Magazine*; but in the Cavendish in the 1930s we were hardly aware of other primary scientific journals in the United Kingdom, and the time had certainly not yet come when the (American) *Physical Review* and *Physical Review Letters* began to attract papers from Europe. The *Zeitschrift für Physik* and *Annalen der Physik* had great prestige, but as far as I remember they published only in German, and the *Zeitschrift* was already coming under criticism for 'accepting everything'—particularly from eastern and south eastern European countries where physics was not yet at a very high level. It was, perhaps, part of a *Drang nach Ostern* in publication policy. For us, the *Proceedings of the Royal Society* were the thing, and it was only after the war that other journals in the United Kingdom became competitive, so far as the atomic physicists were concerned.

Sir Arthur Eddington at that time was well known, particularly for his popularisation of Einstein's work in *Space, Time and Gravitation*. I remember a reference in some newspaper to 'The new physics of Eddington and Jeans', and as both were popularisers in physics this struck the pupils of Rutherford and Bohr as rather odd. Eddington published a paper in 1928 in which a relativistic wave equation for *two* electrons was derived, and according to his theory, based on Dirac's spin matrices, the fine-structure constant $2\pi e^2/hc$ had to have the value 1/136. I remember Eddington presenting this work at the Royal Society, and Rutherford's indignant intervention that the true number was nearer 1/137 and Eddington saying 'well, perhaps it should be 136 from the matrices and one for the teapot'; and indeed in a later paper he derived 137. I was more interested in the fact that Eddington's equation contained two times, t_1 and t_2, one for each electron. I could not imagine what this meant, and went up to the observatory to discuss it with him. I don't think he appreciated that interpretation was a difficulty, and this was curious because my recollection is that, in those days, everyone in Cambridge accepted quantum mechanics at once, together with the probability interpretation which necessitated a single time in the Schrödinger equation. Eddington was clearly one of the great men of those days who worked alone, without constant interaction with his peers. Perhaps such men hardly exist any more in the field of physical science.

Peter Kapitza played a great part in the Cavendish at that time, although his work in the Mond Laboratory was not in the mainstream of nuclear research. He was interested in everything that went on and ran a weekly meeting—the Kapitza Club—where we discussed whatever was new. Kapitza presided in his fluent but almost incomprehensible English. He used frequently to visit the USSR to see his family and scientific friends and used to boast that he was the only citizen of his country with his visa marked for unlimited journeys in and out. But as everyone knows, in 1934 he went to Leningrad, and could not return, his wife alone going back to Cambridge to sell their home in Storeys Way and bring the children to Russia. My wife and I also went to Leningrad at that time; we had planned to go with the Kapitzas, by car to Bergen and round the Gulf of Finland, but in the end there wasn't room in the car and we went by boat from London. The occasion was a meeting to commemorate the centenary of the birth of Mendelyev. I believe I got an invitation because I had helped J Frenkel to prepare his book on liquids for Fowler's Oxford series. Anyhow I met Frenkel in Leningrad, and my most vivid recollection is of returning with him to his house after a visit to the Ioffé Institute, with the promise of lunch at 4 PM—and his going upstairs to have a word with his wife and coming down with a beaming face and saying: 'I was wrong about the time, my wife says lunch is at six, and this is fine because it gives me two hours to explain my latest theory'. His son, many years afterwards, was delighted by this story.

1932 was at the height of the great depression and a year when many people foresaw the regime of Hitler. If it was the *annus mirabilis* for the Cavendish, it was not for most other parts of society. I think the young people at the Cavendish were very conscious of what was happening and as concerned about it as anyone else—but I have no personal recollection of any effect on the work of the laboratory. At that time I think we had visits from German colleagues who a year later had to leave Germany for good; but whether they, or any of us, really appreciated what was going to happen, I do not know. Of course, as far as I know, no one dreamed of the consequences of the discovery of the neutron.

In 1932 Professor Arthur Tyndall, who had recently built the new physics building at Bristol with funds from the Wills family, asked me to accept the Chair of Theoretical Physics at Bristol to succeed J E Lennard-Jones, who was coming to Cambridge to a Chair of Theoretical Chemistry. Cambridge was so much the centre of things that I was uncertain whether to accept. I first asked Rutherford's advice; 'Look at me', he said, 'I went to McGill and Manchester, and came back. Of course you should go!' I then asked Fowler and he advised me to go too. He said 'your opportunities for research at Bristol will be much better than they will *grow to be* in Cambridge.' At that time I had six lectures a week, three for Part II maths and three on quantum theory in the Cavendish, together with eight hours of supervision at Caius. I do not remember this as an enormous burden although I found problems for the Mathematics Tripos increasingly hard to solve; but after all it was only for twenty weeks each year. But Fowler envisaged, I am sure, the increasing claims that a Cambridge college made at that time on its Fellows for committees and pastoral work, and, although I could have resisted them, I would not have been happy in doing so. I remember too in later years feeling a certain indignation as my more promising former students became increasingly involved in such college work.

In any case, I am immensely grateful to Fowler. I don't think I ever learned any physics from him after his lectures on statistical mechanics, excepting of course his hint about collisions between alpha-particles, but there he was, in a way above the battle, doing the right things, giving the right advice, the true father of theoretical physics in Cavendish. It was a tragedy that a long-standing illness and overwork during the war lead to his early death in 1945.

He was, before his illness, a man of very fine physique, with a half-blue for golf and a fine performer in most other games. In the First World War he fought and was severely wounded in Gallipoli. He could speak his mind magnificently and could be downright rude. I remember his saying (but not in Lindemann's presence) 'it is an impertinence of Lindemann to write a book on quantum theory'. Once, at a lecture by Harold Jeffreys, perhaps at the Cambridge Philosophical Society, he said 'I've been asleep, but I know what Jeffreys has been saying is all wrong. . .'. At the same time, as E A Milne wrote in the obituary for the Royal Society, nobody minded and what we remember is his staunch friendship and disciplined optimism. He was not perhaps a genius, but he was supremely able and successful in whatever he touched, and the Cavendish of 1932 owes more to him than is, perhaps, widely recognised.

Notes

1   Wilson A H 1980 *Proc. R. Soc.* A **371** 1
2   Mott N F and Jackson J M 1932 *Proc. R. Soc.* A **137** 703
3   Hulme H R 1932 *Proc. R. Soc.* A **138** 643
4   Ellis C D and Mott N F 1933 *Proc. R. Soc.* A **139** 369
5   Ellis C D and Mott N F 1933 *Proc. R. Soc.* A **141** 502
6   Fowler R H 1936 *Statistical Mechanics* 2nd edn (Cambridge: Cambridge University Press) 870pp

3.3 The Development of Electrical Counting Methods in the Cavendish

W B Lewis

In 1929–30 I was coming through my final undergraduate year and hopefully had put myself down for research in physics but 'in anything other than radioactivity'. A month or two later the Professor—Rutherford—called me to his office and said 'I am told you understand about these wireless valves. We are just beginning to use them in our alpha-ray work. If you get through your exams all right I would like you to join our group'. In modern words electronic circuit engineering or electronics was my qualification of special value, but I soon found myself immersed in radioactivity as well; at that time, however, 'active hard valve switching circuits'—to use the old language—for 'vacuum tube bistable switches' were hardly known and I had my moment of frontier ecstasy a few years later when I had the first vacuum tube scaler counting nuclear particles. In retrospect that looks like no great leap for mankind but one minute step that was paralleled by many almost identical steps elsewhere as circuit electronics was spreading. Those controlled electronic switches were, however, basic to the major advances that electronics brought to the computer.

The well known photograph 'Talk Softly Please' (figure 3.3.1) by C E Wynn-Williams in 1932 can serve to illustrate many aspects of counting technique of that period. The site is one of the larger laboratories into which the Cavendish had been able to expand by the exodus of the engineering faculty to new buildings off the Trumpington Road. The room had been the 'engineering-drawing-office'. Then as a research laboratory it provided space for L H Gray and G T P Tarrant working with gamma-rays, for H Carmichael working on ionisation bursts in the cosmic rays, for W E Duncanson, H Miller and A N May on alpha-ray induced nuclear disintegrations and A I Leipunsky hunting for neutrinos (later he headed the USSR fast nuclear reactor programme). The door shown open on the left led into Chadwick's laboratory where he discovered the neutron in the same year, 1932. Rutherford (with a cigarette) was going his rounds through the laboratory and had been intercepted by Ratcliffe (back to the camera). Wynn-Williams

Figure 3.3.1 'Talk softly please'. The sign was built by Vivian Bowden and was necessary because the lead from the ionisation chamber to the input valve of the linear amplifier was microphonically sensitive.

switched on the illuminated sign normally 'on' only during counting, and was rewarded by the renowned snapshot. At the bottom right in the foreground is Rutherford's bun magnet which in 1932–3 finally resolved the problem of the long range alpha-ray groups from radium-C' and thorium-C'. A long tube rose vertically from the counting ionisation chamber with a central wire connecting to the DEV tube magnetically shielded in an iron pot, made from a bottle used commercially for mercury, with an added lid sawn from soft iron bar stock. One of the screened leads connecting to the amplifier trolley carried the output line from the DEV, another shielded twin carried the filament power and the third carried the 'High Voltage' for the ionisation chamber. The lead from the chamber to the DEV was microphonically sensitive, and that was the reason behind the signal light. That brings us now to the extremely large amplifier and its three batteries of power supplies. It was in fact the original linear amplifier of Wynn-Williams and Ward that had been completely dismantled and reassembled by Wynn-Williams and myself in the summer of 1930 because of intermittent 'bad contact' faults (many of them internal to the components) that from time to time introduced spurious counts. It was mounted on the heavy castors shown to take the weight of the batteries and allow the precious heart of the system to be moved from one laboratory to another. Various switches and rheostats were included for the operations of recharging the batteries. There was also a milliammeter for indicating the anode current of the DEV. Since the grid of this valve was 'floating' the anode current and the amplification were influenced by strong sources of gamma-rays. The milliammeter was not really needed for work with the bun magnet but was important in earlier work on the ranges in air of the long range alpha-rays.

In the early days the output from the amplifier was taken to an electromagnetic oscillograph and recorded on photographic paper tape. This allowed each individual count to be assessed for interfering spurious counts, including microphonics, and also for superposition of pulses where the amplitude was significant. In the 'drawing-office' however counts were made automatically by the three-stage 'scale-of-two' thyratron counter introduced by Wynn-Williams together with automatic timing and recording devices. For certain purposes for recognising and eliminating microphonics a magnetic tape recorder of the early Blattnerphone type installed at the end of the laboratory was sometimes used.

One of the earliest and most significant achievements of the linear amplifier technique was that the sensitivity was high enough to obtain an identifiable response from as little as about 2% of a single alpha-particle track or from a few millimetres of the track of a proton in air at atmospheric pressure. This sensitivity was exploited most ingeniously by Sir Ernest Rutherford, F A B Ward and C E Wynn-Williams in the 'differential' ionisation chamber. It is of interest to note that their paper, *A new method of analysis of groups of alpha-rays, (1). The alpha-rays from radium-C, thorium-C and actinium-C*[1] included two plates showing examples of the oscillograph records. It was received on August 15 1930 and published on September 3 1930. During this same interval the linear amplifier was stripped and rebuilt as already mentioned. Within a year it gave rise to two further papers.[2,3] After one more paper still exploiting the resolution of the differential chamber on the alpha-particles from the radioactive emanations and 'A' products and from polonium, the amplifier was moved to the larger laboratory to work with the annular magnet as shown. The above papers all included a good discussion of the techniques as they developed. My book on *Electrical Counting* even includes mention of the famous 'sealing-wax'.[4]

> The most satisfactory means of connecting the collecting electrode to the first valve is probably by a thin wire along the axis of a wide metal tube. This keeps the capacity low but provides a large volume from which ions may be collected. To avoid this collection of ions the wire should be at the same potential as the surrounding tube. Where a difference of potential of only a volt or so occurs it has been found satisfactory to run the wire through a piece of narrow quill glass tubing along the axis of the wide metal tube. The insulation at the ends of the tube may be improved with sealing wax to reduce the leakage of charge from the glass to the collecting electrode. It is presumed that the outer surface of the glass tube quickly takes up such a potential that there is no further collection of charge. The quill tube by preventing movement of the wire also minimises microphonic effects.

Notes

1   Rutherford E, Ward F A B and Wynn-Williams C E 1930 *Proc. R. Soc.* A **129** 211
2   Lord Rutherford, Ward F A B and Lewis W B 1931 *Proc. R. Soc.* A **131** 684
3   Lord Rutherford, Wynn-Williams C E and Lewis W B 1931 *Proc. R. Soc.* A **133** 351
4   Lewis W B 1943 *Electrical Counting* (Cambridge: Cambridge University Press) p. 7

3.4   The Basic Improbability of Nuclear Physics†

Lord Bowden

I should like to discuss the extraordinary effect Rutherford had on the whole subject of nuclear physics, and how Rutherford's great impact depended on chance. The fundamental work he did all involved the use of alpha-particles as energetic projectiles with which he could probe matter, foils of gold or other metals into which the particles went, and little screens of zinc sulphide upon which the scattered particles fell and produced flashes of light which could be observed through a microscope.

Now this is all very familiar, but what is perhaps not so familiar is the fact that the substance that Rutherford used was always the same. It was zinc sulphide to which had been added about one part in ten thousand of copper. This combination, quite uniquely and quite unexpectedly, was enormously efficient in the way in which it transformed the energy of the incident alpha-particle into a flash of light. About a quarter of all the incident energy was converted into light in the first place, and of that quarter almost all came out in radiation to which the eye is most sensitive, in the yellow–green area of the spectrum. So Rutherford had a quite extraordinarily sensitive method of detecting alpha-particles and he could detect them when they had been slowed down so that their residual range was only a few millimetres of air. (They came off having energies corresponding to a range of perhaps 7 cm of air.) So the flashes were quite easily seen, and with the primitive apparatus that he had Rutherford was able quite unambiguously to prove that matter consists of a sort of planetary system having a nucleus at the middle and clouds of electrons surrounding it.

Now had it not been for the zinc sulphide Rutherford would never have been able to do this work, and this is a very remarkable point. I do not think people realise just how incredibly improbable it is that there should have been, just at the time Rutherford was looking for it, a particular crystal

† This paper is based on a talk delivered at the conference on The Neutron and its Applications, held in Cambridge in September 1982 to commemorate the 50th anniversary of the discovery of the neutron.

which had this peculiar and totally unexpected property. Think for a moment of what would have happened if Rutherford had not bombarded gold with alpha-particles and seen the little flashes they produced. The flashes, perhaps a few hundred of them, seen by himself, by Geiger, and by young Marsden who was his research student, really created nuclear physics, because they established the existence of the nucleus and made it evident that it existed and ultimately would be able to be studied. Had this not happened one likes to speculate what might have been the effect, not only on physics but on the modern world. The achievements of Rutherford, I am sure, would not have been rivalled by anyone else for at least ten or fifteen years. If nuclear physics had not got started then, and had gone on at the usual sort of rate, we should never have arrived at the possibility of the nuclear bomb at the beginning of the Second World War. It was this of course that transformed the nature of nuclear physics from being an esoteric subject of academic study to being an economic industry.

The point I wish to make is that the original work which Rutherford did was made possible only by this very curious substance [zinc sulphide] and the same is true of all his later work. Not only did he detect the scattered alpha-particles that led him to the nuclear theory, but he also found that it was possible to transmute nitrogen and produce long-range protons by bombarding nitrogen in a gas chamber. When Cockcroft and Walton produced the first artificial disintegration of lithium, they knew that they had done it because they saw alpha-particles flashes on zinc sulphide screens.

The next improbability is this one: that C T R Wilson, as a young man, was very fond of climbing mountains. He continued climbing mountains all his life and I believe I am right in saying that when C T R was ninety his family asked him what he would like for his birthday present and he said another pair of climbing boots. C T R was very interested in the way in which great big clouds formed and little drops of water were produced when the air rose and thus cooled itself and condensed. The process of cloud formation was one he observed when climbing. He was able to deduce that if the air expanded in the ratio of 1.25:1, then negative ions would act as foci on which water would condense. If the expansion rose to 1.3:1 then positive ions would do the same and if it were as high as 1.4:1 then any molecule could form a centre for a drop of water. By keeping the expansion ratio within the range of about 1.3:1 to 1.35:1 it was possible to arrange that when the expansion occurred alpha-particle tracks were made to be visible.

Now C T R deduced all this and then he decided to make a chamber, and Rutherford used to tell a story that he started to make the chamber when Rutherford was about to go on a trip to New Zealand, and that when he came back he found C T R still grinding the same piston into the same cylinder. But C T R was a man of enormous pertinacity and he produced it. Again one feels, though, that it was a fairly improbable event that anyone watching clouds and looking at the haloes which are produced round the sun by drops of water should end up by producing a method by which the alpha-particles can so easily be detected.

The third improbability is just as incongruous. People decided that the thing to do was to try and detect alpha-particles by measuring the few ions they produced with a valve amplifier. So the problem arose as to how to make a valve amplifier which didn't itself generate spurious pulses as big as those that were being looked for. Now the modern generation has really no idea how difficult it was to get components that were as good as they were said to be. I remember that the first amplifier I tried had paper condensers connecting the valves, and these produced spurious pulses of their own. I deduced, probably wrongly, that a good cosmic ray burst in a paper condenser would produce the equivalent of a spurious pulse, so that one could count cosmic rays without having the alpha-particles going anywhere near the ionisation chamber at all. This was not what one was trying to do. It was then discovered that by using mica condensers things weren't so bad, but then the problem arose of the first valves. If you used ordinary valves you found that they were always emitting large numbers of little pulses. Bits of thorium, for example, got off the filament, found the grid, and would then emit the odd pulse now and again. However, round about this time the Marconi Company decided to make a special type of valve to produce radio frequency amplifiers. To do this they wanted to reduce the capacitance between the grid wires and the anode wires, so instead of bringing them down to the same base they made a valve about two inches across and brought the anode out at one side and the grid out at the other. This was the so-called DEV valve, and Wynn-Williams discovered that by underrunning it at about half the temperature its filament was designed for it was virtually free of spurious pulses. Wynn-Williams was therefore able to make amplifiers, which everybody used, and which everybody takes for granted today, which worked perfectly satisfactorily, were the first of their kind to be built, and whose quality couldn't be reproduced for a long time until people discovered other valves that were as good.

So we have these three remarkable and unexpected phenomena which made nuclear physics possible, which made the development of the whole work of the Cavendish possible, and which in turn led to the explosively rapid development of modern physics, nuclear power, and of course atomic weapons as a result of the work done in the war. Although we are apt to take it very much for granted today, it might never have happened. Rutherford was not a man to waste his time on a project which did not actually work. He knew that you could get scintillations counted by using small diamonds, but they were not easy to get, they were expensive, and they were not anything like as efficient as the zinc sulphide. So the whole of nuclear physics and the whole of our modern world to an extraordinary extent depends, I believe, on the fact that if you take zinc sulphide with one part in ten thousand of copper and make crystals of an appropriate size, they are quite abnormally sensitive to alpha-particles and produce bright flashes of light which can easily be seen. I find it wonderful to think of three men in a dark room in Manchester looking at a small piece of zinc sulphide which was covered in flashes of light, and it was from these flashes of light that they deduced the fact that the ordinary atom consists of a planetary structure of electrons with a very small nucleus in the centre.

Rutherford was a man who would always think of something to do. Before he came to Cambridge he had invented a system of wireless telegraphy which was better than anything Marconi had at the time. I'm quite certain that if he had not had this alpha-particle detector he would have chosen something quite different to do, and would have become famous for a quite different subject. What our world would have been like had this happened, the world only knows.

3.5   The Scale-of-Two Counter†

C E Wynn-Williams

In my talk this evening I would like to take you back in time to the Cavendish of 25 or 30 years ago, and tell you how first the thyratron ring counter, and then its successor, the scale-of-two counter, came into being during the development of the new technique of high-speed counting. Those were still the days of sealing wax and string, when we had not even an electric soldering iron or a simple cathode-ray oscilloscope, and we had to make out a reasonable case for going out to the local radio shops to buy a few simple electronic components.

Prior to 1928, experiments involving the counting of individual alpha-particles were carried out at the Cavendish either by the time-honoured but slow method of visual counting of scintillations or by counting the kicks of a string electrometer connected directly, without amplifiers, to a Geiger counter. In 1928, however, there was a change. A year or two previously, Greinacher had succeeded in detecting individual alpha-particles and protons by amplifying linearly the ionisation currents from a few millimetres of their tracks in air. In Cambridge, F A B Ward and I were specially interested in this new method of counting, and in 1928–9 in collaboration with H M Cave, we investigated its possibilities by using it in an experimental determination of the rate of emission of alpha-particles from radium, which involved counting over 10^5 string galvanometer deflections recorded on film.

The success of this trial was most encouraging. During the subsequent months, considerable improvements were made to our amplifier, and the string galvanometer was replaced by a faster home-made moving-iron oscillograph. I think Duddell himself might not have been displeased with this, since it used his method of attaining high-frequency response by means of a taut wire loop. The technique was gradually improved until not only could alpha-particles be counted at high speed even against a strong disturbing background of gamma-radiation, but also it was possible to distinguish, by their different sized deflections, the alpha-particles of different energy groups.

† This paper is an edited version of Wynn-Williams's Duddell Medal Address, printed in full in the *Physical Society Yearbook* (1957) pp 53–60.

Using this equipment, Lord Rutherford, assisted by Ward and myself, and joined later on by W B Lewis, investigated the long- and short-range alpha-particle groups emitted from various substances. During the same period, Sir James Chadwick, assisted by J E R Constable and E C Pollard, were successfully applying the new technique to the problems connected with artificial disintegration by alpha-particles.

One great disadvantage of the new method of counting was that one did not know the result of an experiment immediately. By the time the film was processed and analysed, a short life active deposit source was well and truly dead, and a fresh one would have to be prepared for the next experiment on a subsequent day. Automatic mechanical counting, either alone, or ancillary to the photographic method of recording, was thus very badly needed at this time. Ward and I had tried to do something about this by linking up with the amplifier and oscillograph one of the Cavendish electromechanical registers used in the old scintillation counting days. It certainly did count alpha-particles, but it was sluggish and unreliable, so we only used it for rough counts prior to photographic recording.

About the middle of 1929 came our introduction to the thyratron. A W Hull, of the American GEC, who was visiting the Cavendish, gave us first-hand information about it which made it clear that here was just the device needed for automatic recording.[1] The thyratron could be tripped by a fleeting positive impulse from an amplifier, even of only a few microseconds' duration. The subsequent large arc current could operate a robust mechanical counting meter, and it would continue to flow until the anode circuit was interrupted by the meter when recording was completed.

While we were still pessimistically wondering how long it would be before we could lay hands on one of these new thyratrons, a Cavendish enthusiast, N A de Bruyne, had opened up an old T.15 transmitting valve, introduced a globule of mercury, evacuated, baked out and sealed off the valve. He proudly presented this to us with the casual remark, 'Here's a thyratron for you'. So, within a few days of the talk with Hull, we were able to verify that a thyratron really could be used for automatically counting alpha-particles. Unfortunately, before de Bruyne's thyratron could be used on a live experiment, it met with a sad accident a day or two later, when lent to another Cavendish enthusiast. In due course, however, the BTH Company kindly presented the Cavendish with some thyratrons and de Bruyne and Webster successfully used single-thyratron counters for automatic recording with their Geiger counters.[2]

Rutherford was very fond of using the single-thyratron counter at his Royal Institution lectures. The changing dial figures, the sharp click of the mechanism and the bright flash of the arc, all helped to convey to the audience that alpha-particles really were being counted. Reporters, however, sometimes missed the point, for one morning, the following statement appeared in a leading newspaper: 'Last night, at the Royal Institution, Lord

Rutherford disintegrated the atom with a blue flash and a noise like rapid machine-gun fire!'

Early in 1930, it occurred to me that the single-thyratron counter did not fully exploit the characteristic properties of a thyratron. Though a fleeting impulse was caught, the thyratron was insensitive to subsequent impulses until the associated mechanical meter had finished registering the impulse. With the modified telephone call counters then used, this dead time was about 1/25 of a second, which is enormous compared with the thyratron deionisation time.

I decided that what would be needed would be a ring of thyratrons operating in the following manner. Initially, there would be an arc in one of them, say thyratron P. Because of this arc in P, the next thyratron, Q, and no other, would by some means be made ready to respond to an incoming impulse. When thyratron Q responded, the arc in Q would, in a similar manner, immediately prepare thyratron R to respond to the *next* impulse to arrive; it would also, by some method, cause the arc in thyratron P to be extinguished. If the connections of all thyratrons were similar, successive impulses would thus step the arc from thyratron to thyratron round the ring. If there were N thyratrons in the ring, and one of them operated a mechanical counting meter, N times the change of meter reading, plus a number less than N determined by the initial and final state of the ring, would equal the total number of impulses counted. This, then, was the principle of the electronic ring counter, the forerunner of the scale-of-two counter.

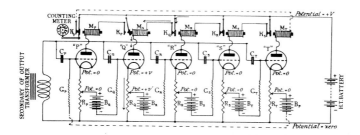

Figure 3.5.1 Redrawn from *Phys. Soc. Yearb.* 1957, p. 55. See text for details.

One simple practical arrangement for achieving this would be the circuit arrangement shown in figure 3.5.1. All thyratrons would be given excess negative grid bias, so that normally none would respond to an applied impulse. An arc in, say, thyratron Q would, however, set up a potential difference in the cathode resistor, which would reduce the grid bias of thyratron R to a normal value, so that R would respond to the next impulse. At the same time, the relay in the circuit of thyratron Q would extinguish the arc in thyratron P by breaking the anode circuit. Suitable inductive lags in the establishment of the arc currents would ensure that one impulse would not fire all the thyratrons in quick succession.

The least number of thyratrons which will work in this manner is three. Three thyratrons and three makeshift relays were therefore suitably wired up, and, after a little coaxing, on 19 March 1930, the lash-up system worked correctly.

Eventually, I succeeded in collecting together sufficient reliable components to make a ring of four thyratrons. The large thyratrons I had were directly heated, and required 7 amperes each. The heating transformer was consequently a massive bit of ironmongery, with the thyratrons soldered directly on to the four thick, air-spaced secondary windings. True Cavendish thrift in the use of copper in the primary winding ensured that the transformer also emitted considerable warmth and needed no pilot warning light, since it buzzed happily on the Cambridge 100 volt 90 c/s [Hz] town supply. However, this ring circuit worked extremely well and was in use for several months. A thing which always intrigued visitors was that, with all this electronic progress, we still reset the arc to the starting position each time by pulling the end of a yard of thread. In those days it would have been considered extravagant frivolity to provide an earthing relay.

In due course, I improved the counter by changing the method of arc extinction. Figure 3.5.2 shows the new arrangement. The relays were removed, and all cathodes were coupled to their neighbours on either side by condensers. With this arrangement, when an arc strikes in, say, thyratron Q, the cathode potential of Q rises suddenly, and a positive voltage impulse is communicated to the cathode of thyratron P. If this is of sufficient duration, it suppresses the arc in P until deionisation is complete, and so resets thyratron P.

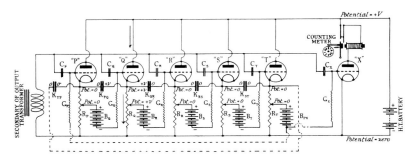

Figure 3.5.2 Redrawn from *Phys. Soc. Yearb.* 1957 p. 56. See text for details.

This change resulted in faster working and made the ring counting action, for the first time, entirely electronic, although of course a mechanical meter still had to be operated once for each complete counting cycle. In this form, the circuit had a resolving time of about 1/600 to 1/700 of a second, and it was used by Rutherford, Lewis, Ward and myself, and by B V Bowden, who joined the team later.[3]

During the months that the ring was being used, it occurred to me that an entirely electronic counter could be made by substituting additional thyratron rings for the mechanical counting meter. A multiple-ring counter could thus be made which would operate according to the decimal scale, or any other desired scale of notation. At the time, it did not seem worthwhile making a decade counter of this type; it would have been clumsy and wasteful of thyratrons, and it would have had no real advantage over the existing apparatus. Consequently, without making experimental tests, the possible future development of multiple-ring counters was merely put forward as a suggestion in an account of the work which was published in July 1931.[4]

Although at least three thyratrons are needed in a ring which employs relays for arc extinction, it seemed to me that with condenser extinction a ring should work satisfactorily with only two thyratrons. Further, the circuit might be simplified, for, with one thyratron normally extinguished and the other carrying an arc, no special circuit arrangement would be necessary to decide which thyratron should next fire; only one thyratron, namely, the extinguished one, *could* fire. This meant that anode resistors could be used instead of cathode resistors, and the cathodes could therefore be at common potential. It also meant that separate heater windings and separate bias batteries would no longer be needed. A quick experimental test showed that this simplified circuit was quite satisfactory, and the diagram shown in figure 3.5.3 which illustrates the changes, was published in the paper I have just mentioned.

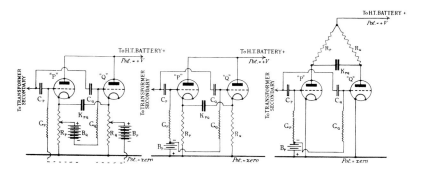

Figure 3.5.3 Redrawn from *Phys. Soc. Yearb.* 1957 p. 57. See text for details.

Looking back now, it amazes me that at this stage I did not immediately think of the scale-of-two counter, and that in the months subsequent to the publication of the paper, no one else forestalled me; for in the one paper I had stated that this simplified two-thyratron ring really worked, and I had suggested the possible development of multiple-thyratron rings. I think the reason why I did not think of it at the time must have been that I believed then that an electronic coupling device would be needed to pass on a positive driving impulse from one ring to the next. I dared not experiment on these lines at the time because we had no spare thyratrons, and it would have been most unwise to interfere with the only ring we had, which was piling up experimental results.

In due course, in October 1931, it occurred to me that in the simplified two-thyratron ring a suitable positive driving impulse was, in fact, always developed at the anode of a thyratron when the arc was extinguished. It was only necessary, therefore, to link one of the anodes, through a simple resistor and condenser circuit, to the grid circuit of the next two-thyratron ring, and so on for any number of rings. Thus was the scale-of-two counter finally devised.

During the next few days, as opportunities occurred, tests were made by quickly converting the four-thyratron ring into a two-stage scale-of-two counter, and subsequently restoring it to the *status quo* for normal counting when required. It was soon clear that the scale-of-two counter worked and was, in every way, superior to the ring counter. It was stable, free from critical adjustment, almost foolproof, and it seemed to have better resolving power. Unlike the ring counter, it was not fussy about its high-tension supply, for it operated perfectly from the Cavendish 200 volt DC mains without any smoothing or stabilising circuits.

This was so encouraging that within a month a three-stage scale-of-two counter had been built, and its resolving time measured and found to be about 1/1250 of a second. The measurement was carried out with a glorious system of rubber bands, workshop nails, the inevitable Cavendish string, resistance wire, wooden laths and a long piece of strong catapult elastic. One stretched the elastic, stood clear, let go and hoped there would be no serious damage as a number of nails were whisked away in quick succession. The same contraption also did its duty at one Royal Society Soirée, demonstrating that the counter really *did* catch trains of up to eight impulses separated by time intervals of the order of 10^{-3} second. Nowadays, costly electronic gear would be provided for such purposes, but this was much more fun.

This counter, the circuit of which is shown in figure 3.5.4, and an identical one which quickly followed it, were in constant use in the Cavendish from that time until after I left in 1935. On a recent visit there, I searched to see if any of this old equipment could be found, in order to make a slide for this talk, but I was unsuccessful. I did, however, find just one counter of the next vintage, complete with thyratrons; some of you may have used it in the Part II class. This counter was one of about six which the workshop made to a specification Lewis and I prepared just before I left. There was a steady demand for this specification, so it was duplicated and, with circuit blueprints, was given to all who asked for it. I still have a copy; the last line in

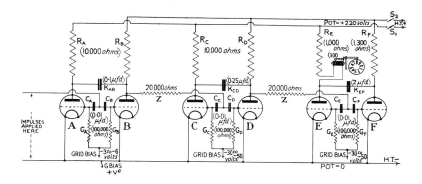

Figure 3.5.4 Redrawn from *Phys. Soc. Yearb.* 1957 p. 58. See text for details.

it makes one think a bit in these inflationary times; it reads 'Total cost, excluding thyratrons, approximately £10'.

Before I left the Cavendish, several refinements were made to the counters. Valve coupling was provided between the first two stages, and Lewis deliberately introduced 'dead time' into the first stage and reduced the resolving time to about 1.5×10^{-4} s. Then we discovered a place where we could buy secondhand relays and uniselectors for a shilling each, so I was really able to enjoy myself devising ancillary equipment for the counters.

Instead of telephone call counters, uniselectors were used as the mechanical counting meters. Their operation time, after suitable treatment, was shortened to between 1/80 and 1/120 of a second. In conjunction with the resolving time of 1.5×10^{-4} s this meant that the system could safely be used for rates up to about 8000 particles per minute, and was capable of working to an accuracy of about 0.5%. The same uniselectors also provided another facility; the eight-times table was wired on the contact banks, and this, with a little more relay circuit arithmetic, multiplied by eight and added in 1, 2 and 4 as required, thus instantly converting the binary scale result of a count to the decimal scale when the counter was switched off.

Other uniselectors, driven by a master clock, timed runs for chosen preset intervals, and also indicated the mean time of each run for source decay correction. Yet another uniselector numbered the runs consecutively. The *pièce de résistance*, however, was a home-made electric typewriter which investigated, in turn, the states of the various uniselectors, and printed on a paper tape the run number, the duration and mean time of the run, a designation and the total number counted. Cavendish thrift still prevailed, for the old type-wheel, which I still have, contains rubber type from a sixpenny toy printing set, and the rolls of paper tape were of various pretty colours, for they all came from twopenny packets of paper streamers.

In a second version of this apparatus at Imperial College, there was also a remote control device, which enabled the whole equipment to be operated by means of a telephone dial over a pair of wires, which also carried the impulses which were to be counted. There was even a digit which, when dialled, caused the entire equipment to shut itself down and switch the power off, but this was a mixed blessing since the digit was sometimes dialled in error. Finally, just before war broke out, a programme device was added, which could control experimental conditions and carry out cycles of pre-arranged runs by remote control of the equipment.

Since the first ring and scale-of-two counters appeared, great improvements have been made by various people. Circuits of both types have been developed for the various kinds of two-state devices, which have very low resolving times. One important advance is that scale-of-two circuits are now usually made to record on the decimal scale by an ingenious principle. When nine impulses have been recorded by a four-stage scale-of-two unit as '1001' in binary notation, a circuit recognises this and arranges that the next impulse which arrives resets this unit to the zero state. Thus, the counting cycle for the unit is changed from 2^4 to 10.

The application of counters now extends beyond nuclear research, for which they were first developed. It is pointless to list their uses, for any event which can be made to produce a suitable voltage impulse can obviously be counted by these instruments. Besides serving as a variety of physical measuring instruments, however, a counter may be incorporated as a circuit element essential for the operation of some larger piece of electronic equipment. I need only mention briefly two examples. One is the electronic computer, which performs its arithmetical and storage operations most conveniently in binary notation, and translates decimal numbers into this scale or vice versa as required. The other example has been very much in the news recently, namely ERNIE, the machine which determines whether or not you get a prize in the Premium Bond scheme. It would have been pleasing to me to be able to tell you that ERNIE incorporated a scale-of-two counter for doing its arithmetic, but this is only partly true. Although there are single binary stages, the counting is done by ring counters using transistors and ferrite core transformers, working according to scales of three, six, ten and twenty-four.

Looking back now, it seems to me that sooner or later someone was bound to have devised a scale-of-two counter. I think it was really good luck more than anything else which resulted in my being the first person to do so. I happened to be at the Cavendish when there was a real need for such a device, the thyratron had just been developed, the BTH Company were able to supply us quickly with thyratrons, and I was very interested in electrical circuits. But for this fortunate combination of circumstances, I am sure it would be some other person, and not me, whom you would be honouring this evening.

Notes

1   Hull A W 1929 *Gen. Electr. Rev.* **32** 390
2   de Bruyne N A and Webster H C 1931 *Proc. Camb. Phil. Soc.* **27** 113
3   Rutherford E, Ward F A B and Wynn-Williams C E 1930 *Proc. R. Soc.* A **129** 211
4   Wynn-Williams C E 1931 *Proc. R. Soc.* A **132** 295; Wynn-Williams C E 1932 *Proc. R. Soc.* A **136** 312
5   Hayward R K and Bubb E L 1957 *P. O. Electr. Eng. J.* **50** (1) 1

3.6 Metropolitan-Vickers Electrical Company and the Cavendish Laboratory

T E Allibone

I have been asked to write about the association of the Metropolitan-Vickers Company with the Cavendish Laboratory, culminating with the great experiment of artificial disintegration in 1932. The story is a very long one involving many people, some of whom are, alas, dead, and in the space I have been allotted only parts of the story can be told. To the best of my knowledge most of the story has not previously been written for publication; I am not aware of all the details nor have I found time to visit Churchill College where Sir John Cockcroft's papers are kept—doubtless some of my omissions could have been filled from that source.

The most direct link is between the director of research of Metropolitan-Vickers, Mr A P M Fleming and Rutherford. At the beginning of World World 1 Fleming was the Superintendent of the Transformer Department of the Company, then called the British Westinghouse Company, a company created by the great American engineer, George Westinghouse in 1899. In that year forty young college-trained engineers had been sent to Pittsburgh for a two-year period to learn about the business of design and manufacture of Westinghouse products; most of these men returned to make their career in Trafford Park and were variously called the 'Holy Forty' or 'The Forty Thieves'. Fleming was one of them and Miles Walker, fresh from having taken the Mechanical Sciences Tripos in Cambridge in 1899 was another. Fleming spent his life with the Company, Miles Walker was appointed in 1907 to the Chair of Electrical Engineering at the Manchester College of Technology but he remained a consultant to the Company all his life.

Fleming was responsible for testing not only the finished transformer but also the ingredients, magnetic sheet steel, solid and liquid insulation and it was from the Transformer Test Department that he laid down the rudiments of a research department which, in due course, became one of the most famous research laboratories in the country. When an anti-submarine committee was formed early in the war he was invited to join Ernest Rutherford,

150

then the Langworthy Professor of Physics at Manchester University, and share in the experimental work conducted on Lake Windermere.

The war was revealing how backward Britain had been in many spheres of industrial research, notably in the supply of scientific instruments and chemical dyes, and a proposal was made that various industrial groups should each sponsor a research laboratory, partially financed by such groups and partially by Government. Fleming had been to America in 1917 to study the organisation of research in several of the large industries and Government laboratories and on his return he endorsed the idea of creating Research Associations supported equally by industry and by Government under a Department of Scientific and Industrial Research. In due course the Government voted the 'million pound fund' and so were born, first, the British Scientific Instrument Research Association, quickly followed by the British Electrical and Allied Industries Research Association, and then many more. Fleming was made Manager of the British Westinghouse Research Department on October 1 of that year and as time went by he added sections dealing with mechanical and metallurgical engineering problems, a section for instrument standardisation, a physics section, and a high-voltage laboratory for work on problems of AC transmission and distribution. He also introduced apprenticeship schemes not unlike those which Westinghouse had initiated in 1899, schemes for schoolboys leaving at 14 to 16 years of age and college apprenticeships for university graduates in electrical and mechanical engineering. Formal tuition was given to the juniors, but for the college apprentice a two-year scheme of rotation of employment was devised—taking into account, as far as possible, any special wishes of the graduate. This scheme involved spending three months in each of eight departments of the Company ranging from, say, the pattern shop and foundry, design, research and sales departments, some 'outside erection' experience, some shop-floor and management experience, so that at the end of his time he had an extremely good idea of the products, practices and management of the Company and was able, in general, to select the vocation best suited to his interests and ability. During this apprenticeship he earned the princely salary of less than £100 a year which, in the early 1920s barely sufficed to pay the landlady and eke out a modest social life in and around Trafford Park, Stretford and North Cheshire. Nevertheless, the college apprenticeship became a much-prized status symbol, greatly sought after by graduates from all British and many overseas universities and, later in life, some of these would be found as chief engineers of various supply undertakings and of electrical companies all over the world. Fleming told me in the 1930s that he had trained every one of the Chief Transformer Engineers of seventeen of our competing companies but he did not regret this; on the contrary , he found a comradeship and a willingness to work together for the benefit of the industry as a whole wherever he went. When, later in life, I became a director of one of the three branches of the Company, by then called Associated Electrical Industries Ltd (AEI), every director, except the lay directors, had been a college apprentice. Incidentally, it is frequently said today that boards of companies ought to contain more technical directors; the AEI had a surfeit of them but that did not stop the Company from heading for the rocks in 1963 and floundering a little later.

I must here mention the change in name and ownership of the Company. Before the end of the Great War the Sheffield armaments firm of Vickers began to consider how its huge steel-making capacity could be profitably used in peacetime and it decided to buy out the American Westinghouse interests in Trafford Park, some of which were already held by the Metropolitan Carriage, Waggon and Finance Corporation. Thus in 1919 the Metropolitan-Vickers Company, wholly British, was formed. (J D Bernal always wrote to me at the Metropolitan Vicars Company, but I was never a member of the Greek Orthodox Church!) I think Westinghouse must have retained some financial interest because for many years the original 'Westinghouse Agreement' enabled the British Company to share in Westinghouse patents, a liaison officer was based in Pittsburgh and Fleming and his staff were always welcomed in the Old Trafford Research Department, as George Westinghouse had named his laboratory, up the hill at Pittsburgh.

John Douglas Cockcroft entered the University of Manchester in 1914 to read for a degree in mathematics with physics and French as subsidiary subjects; some of his physics lectures were given by Professor Rutherford just back from the Antipodes. He passed the Intermediate BSc examination a year later and was then caught up by the war, serving in the Signals, mostly in France, for three years. After demobilisation he returned with an ex-serviceman's award to study a subject closer to his war training, a degree in electrical engineering at the Manchester College of Technology under Professor Miles Walker, and graduated in 1920.

Then Miles Walker strongly recommended him to join Metropolitan-Vickers as a college apprentice. After some months in various parts of the Company he elected to spend the rest of the two years in the research department, and indeed he was able to submit some of his work to the Manchester College of Technology and to gain an MSc Tech. Fleming always had very liberal views about the publication of scientific and technological material emanating from one's work in the department; he encouraged staff to publish as much as possible providing it did not reveal any special information of a commercial character likely to be detrimental to the Company's interests. John was a good mathematician, and Miles Walker suggested he should try to go to Cambridge and take the Mathematical Tripos. He applied for, and was awarded, a Sizarship at St John's College and also a Miles Walker Studentship of the Manchester College of Technology, and in addition, Mr Fleming gave him a grant of £50 a year. In view of my experience and that of several of my colleagues, I should think that, in giving this small grant, Fleming would ask him to keep in touch with the

Company and tell him that his travel expenses to and from the works would always be paid. Fleming was a generous man, very interested in young people and always keen to help them in their careers, as indeed George Westinghouse had helped him and his friends.

It will be recalled that, in 1919, at Manchester University, Rutherford had transmuted the nitrogen nucleus by alpha-particle bombardment. Later that year he was appointed to the Cavendish Chair of Experimental Physics in Cambridge, succeeding his former professor, Sir J J Thomson who had been appointed by the Crown to the Mastership of Trinity College. Rutherford invited James Chadwick to join him in Cambridge. 'Chad' had worked as a research student under Rutherford in Manchester, had then been caught up by the war when working with Geiger in Berlin and had spent the four years in an internment camp at Ruhleben, not languishing, but trying to do some scientific research and also teaching one RMA cadet, Charles Drummond Ellis, some physics. Chadwick was awarded a Wollaston Studentship at Gonville and Caius College, Cambridge, in 1919 and collaborated with Rutherford in experiments which produced transmutations of the nuclei of many elements. A year later, a young Russian physicist, Peter Kapitza, asked Rutherford if he could work in the Cavendish Laboratory but was informed that the Laboratory was full, with over 30 research students: Kapitza asked the Professor what accuracy he claimed for his alpha-particle experiments and was given the reply 'a few per cent'; 'Well', said Kapitza, 'you could accept me without noticing any change in the number of your students'! Kapitza wanted to produce intense magnetic fields of perhaps 100 000 to 1 million gauss but I never heard whether this was in the expectation that nuclear changes might be caused by such fields. Certainly, a cardinal feature of Rutherford's career was to 'try it, and see', a philosophy which yielded immense dividends, and throughout his life he viewed the theoretical approach with scepticism, almost, at times, with disdain. (To Fermi in 1934, after Fermi had announced the great increase in neutron absorption as the velocity of the neutron was reduced, he wrote, not his congratulations on the wonderful experiment, but 'I am glad to see you have successfully escaped from theoretical physics'.)

It was into this heady atmosphere of the new Cavendish that Cockcroft arrived in the autumn of 1922 with a letter of introduction to Rutherford from Miles Walker (who was elected an FRS in 1931). Rutherford received him kindly and encouraged him to spend as much of his time as he could afford from his Tripos studies in the advanced practical classes in physics run by Appleton and Thirkill. This he did during the two years 1922–4. As a Manchester graduate he could take a Cambridge BA in two rather than three years, and finished a B* Wrangler, with a Foundation Scholarship at St John's and a State Scholarship to support him for a three-year PhD period in the Cavendish. Once more, Fleming offered him further financial support and encouraged him to keep in close touch with Trafford Park. It was during this period that he examined a very important matter for the Turbo-alternator Department, the penetration of magnetic flux into slots in punchings made of magnetic sheet steel, a paper which has well stood the test of time. For his doctorate, he elected to work on a problem concerning the deposition of thin metallic films onto surfaces cooled to low temperatures, but almost immediately, at Rutherford's request, he became involved in some of Kapitza's problems.

To create huge magnetic fields Kapitza had assembled a very large accumulator and had devised a switch to make and break the circuit, shorting the accumulator through a solenoid of very low resistance; the work is described in two papers in the *Proceedings of the Royal Society* for 1924. By the time I arrived in the Cavendish two years later dozens of large thin lead plates lay scattered in odd corners of the room to which I was directed. The work had been supported by a grant from the DSIR but Kapitza was not satisfied with this source of current and either he, or Cockcroft, or someone else suggested short-circuiting a large alternator for one half-cycle, thus gaining energy from the loss of momentum of the heavy rotor during the short. Miles Walker, a specialist in alternator design, was consulted, as was also the Chief Engineer of the Turbo-alternator Division of the Works.

Here I must bring into the picture Fleming's right-hand man, George McKerrow, a Cambridge Mechanical Sciences Tripos graduate, a linguist, nephew (I think) of the Chairman of the Company, Sir Philip Nash, wealthy, and a fine example—rare in those days or even these days too—of a man embracing 'the two cultures'. His job was 'scientific liaison' and he seemed to know most of the leading scientists in the universities and in business. He was the catalyst in many reactions and one of his jobs was to keep in touch with all the young men Fleming had sponsored here and there, Cockcroft in Cambridge, A J Bradley (FRS 1939) whom Fleming had appointed to the Research Department in the early 1920s to work for the most part directly under Professor Bragg in Manchester, and others. He became a good friend of Kapitza's and when the DSIR awarded the contract for the 1500 kW alternator to Metropolitan-Vickers, it was McKerrow who ensured complete harmony between the three parties; at the end of Kapitza's Royal Society paper[1] describing the generator of his high magnetic fields you will see his expression of thanks to McKerrow for his great help. The alternator had specially reinforced end-windings to withstand the great forces generated during the shortcircuit. The complicated switch, mounted on the shaft, to close and open the circuit at 'current zero' was of Kapitza's design, but John Cockcroft, working in a room only a few paces from the magnetic laboratory, as it was called, undertook very many of the jobs associated with the planning and installation of the biggest venture ever seen—up to then—in the Cavendish Laboratory. I have never come across a full account of all that had to be done, but when I arrived in the Laboratory and worked in the same room as John, he was in and out of Kapitza's laboratory several times *every* day, whilst trying to do his own research. He designed the solenoids through which the large currents passed and I do not recall ever hearing of any failure of these under the tremendous forces to which they were subjected; they are fully described in his long paper in the *Philosophical Transactions* of 1928. At some stage a committee was set up headed by Tizard, the Secretary of the DSIR, to supervise the expenditure on the project, and John was appointed, at a small salary, as its secretary; but he devoted far more time to the magnetic laboratory than corresponded to the small emolument he received. By 1926 the machine was running to schedule and Kapitza's experiments were in progress but there seemed to be many aspects of the work which still fell upon John's willing shoulders.

It is here that I must inevitably come into the picture. I had always wanted to go to Cambridge to study physics and mathematics but I happened to win the Ezra Hounsfield Linley open scholarship to Sheffield University, a scholarship worth far more than the Cambridge college scholarships and my parents recommended me to accept this and not spend another year in school hoping that I might gain a Cambridge award. One of the lecturers in the physics department, Mr J R Clarke, had been the section leader of the Physics Section in Fleming's new department, and it was on his recommendation that I was invited to spend the Christmas holiday of 1923–4 in Trafford Park working with Mr F B Burch, a Cambridge ex-college apprentice. In the spring of 1924, before I had graduated, I was offered a staff appointment, not a college apprenticeship, provided I got a first-class honours degree; to my great surprise I was offered almost twice the apprenticeship salary, by then running at £125 per annum. Mr Fleming gave me a job which might have taken me to work for a year under Professor G von Hevesy, in Copenhagen but which, in the event, took me to the Metallurgical Department of Sheffield University. I was actually allowed to register for a PhD in that department and apart from spending months back in Trafford Park, I was kept in touch with the Company by frequent visits of Mr McKerrow, Major Buckley (Mr Clarke's successor) and others from Manchester. At the same time I became the secretary of Sheffield University Physics Society and in November 1925 we had a lecture from Mr C D Ellis, a Fellow of Trinity, on the work of the Cavendish, especially describing the Rutherford–Chadwick transmutations of elements with alpha-particles. The physics professor, S R Milner, kindly invited me to dinner to meet Ellis and I took the opportunity of asking if there were any scholarships available for non-graduates of Cambridge to study for the higher degree. He told me of the Trinity scholarship and said there were others to be found in the Year Book. I spent the Christmas holidays in Trafford Park and told Mr Fleming of my conversation with Ellis and my intention of applying for one of the research scholarships and he at once said he would supplement any grant that I might get if I would keep in touch with Metropolitan-Vickers. I had found that Gonville and Caius College offered a Wollaston Studentship and, by good fortune, George McKerrow was a Caian, as had been two other members of the Research Department, F P Burch and his brother, C R B, of whom more anon. McKerrow wrote to Chadwick about me and Chadwick said I would have to submit a programme of research to Rutherford and seek admission to the Cavendish. Now my great friend of those days was Brian L Goodlet, then in charge of the High-voltage Laboratory in Trafford Park, later to become Chief Engineer at Harwell and author of the design of PIPPA, the carbon dioxide cooled graphite-moderated reactor, the basis of all the British nuclear power reactors. Goodlet, a British subject, born in Russia, had shot his way down the Nevesky Prospect in St Petersburg during the revolution, had escaped from Kronstadt, and finished up in Sheffield where, unknown to me, he had been taught mathematics by my father and engineering by the man who, in due course, was to become my father-in-law. I had been in the High-voltage Laboratory frequently and felt quite at home with half-million volt experiments. I discussed my hopes with him, telling him I would like to accelerate charged particles in a vacuum tube and bring them out through a window, and see if I could produce any artificial disintegrations. He said the cheapest, lightest and smallest source of high voltages was a Tesla transformer which, operating at around 50 000 Hz had no iron core. So I wrote a proposal to be sent to Rutherford—I think McKerrow sent it via Chadwick but I have lost my copy and cannot now recall how it went to Cambridge; it was based on a Tesla transformer and a vacuum tube which I thought I could make, for indeed, I was a moderately good glass-blower. Again, Fleming said if I could get a grant and be accepted by Rutherford he would help to finance the apparatus. George McKerrow saw Rutherford on my behalf and I was invited to meet him in the Cavendish on March 29 1926. It so happened that Rutherford was giving a Friday Evening Discourse at the Royal Institution on March 26 and McKerrow recommended me to go and hear the great man—'better for the dog to see the rabbit beforehand' he said! This I did so that I was quite familiar with the face, the booming voice and the mannerisms when I was ushered into his small office that Monday afternoon. He had read my proposal and after asking many questions he took me down to the largest and highest room in the Cavendish, a room perhaps 14 to 15 feet high. There I met for the first time John Cockcroft, though Mr McKerrow had often spoken to me about him, and Leslie Martin, the Australian. 'How many volts could you generate in here?' barked Rutherford; I replied 'at least 500 kV but not much more'. Some weeks later he wrote to accept me provided I could be self-financing, though he offered to apply to DSIR for a grant of £100. My application to Caius was successful, I heard in July that the Wollaston Scholarship had been awarded to me and in late autumn I handed over my metallurgical work in Sheffield to Charles Sykes who had taken his physics degree in Sheffield one year after me and had likewise been appointed by Fleming to work in Sheffield first and then in Trafford Park (he was later to be elected an FRS in 1943 for his work in the Physics Section of the Department).

My thesis was to be on the acceleration of charged particles to high voltages, and Rutherford suggested I should accelerate electrons, as positive ions were not readily produced and unless they could be made to rival eight million volt alpha-particles they had no chance of entering the nucleus. During 1927 the windings of the Tesla transformer were made in the Works. I assembled them and made all the ancillary apparatus and the first rudiments of vacuum tubes, but equipment and raw materials were very scarce in the Cavendish of those days and getting anything out of Lincoln, the rather fierce-looking storekeeper in charge of the workshop, was like trying to get blood out of a stone—a story which has been told many times by my contemporaries—and progress was painfully slow. Brian Goodlet came frequently to give advice about the parts for the Tesla transformer and with Cockcroft we dined either in Caius or St John's, being joined from time to time by Mr Fleming, George McKerrow, Major Buckley, Sykes or others of the Company who gravitated to Cambridge.

Leslie Martin left England and his place in our room was taken by E T S Walton from Trinity College, Dublin. I think it was Rutherford who suggested to him that he should try to accelerate electrons by some indirect methods; firstly by accelerating them in a circular orbit (not unlike Sir J J Thomson's electrodeless discharge tube with which J J was working in a nearby room in the Cavendish) but one through which a changing magnetic field threaded perpendicular to the plane of the glass torus in which the electrons would be accelerated by the induced electric field, and secondly, by a linear acceleration in which the particles, in this case positively charged particles, not electrons, would be repeatedly accelerated by a relatively small voltage. In the event both efforts failed though Walton had devoted great skill and determination to his work. The first failed because, to be successful, the electrons must be continuously focused if they are to remain in the correct orbit as their speed increases and Walton had not worked out the focusing equation, whilst the second failed again because the beam of positive ions was not focused as it passed from one hollow cylindrical electrode to the next. Both were stout efforts and needed Walton's great manual dexterity—which was very considerable, he could repair watches—to make the devices work at all with the totally inadequate facilities available in those days. One technique we both acquired with distinction; we kept an eye on John Cockcroft's apparatus and saved it times without number from total collapse. John would come in early in the morning—well, not too early—switch on pumps etcetera for his metal vapour deposition apparatus and then dash out to Kapitza's laboratory, or to St John's or elsewhere, completely forgetful of the need to turn on the water or something else, and one or other of us would find his apparatus in a critical state just before it disintegrated. But he got his PhD in 1927 and the Clerk Maxwell Scholarship to boot!

One more event in the Metropolitan-Vickers Research Department at the end of 1927 and into 1928 had a profound effect on us, on the Cavendish and, in due time, on the world at large; C R Burch in our Physics Section produced 'Apiezon oil'. The bare bones of this story can be found in a paper which Rutherford communicated to the Royal Society in January 1929 but I do not know where else it can be found with the background story, so I would like to include parts of it here because Apiezon oil and Apiezon greases played an essential part in my and in Cockcroft and Walton's research. The oil ultimately replaced mercury in diffusion pumps, the very first of which were used by me after the Trafford Park experiments, later they were used by C and W [Cockcroft and Walton] and all in good time reduced the amount of liquid air which Lincoln made every Monday morning. In the mid-1930s they were used in the forty large transmitting valves in the CH transmitters built around our coasts which gave the early radar warning information on which the Battle of Britain was won. Later, in Oakridge, the uranium isotope separation was made possible by some hundreds of thirty-inch oil-diffusion pumps all using Burch's Apiezon oil. Transformer insulation, paper and pressboard, had always been impregnated with transformer oil in very large vacuum chambers, a completed 132 kV power transformer core and windings would be wholly immersed for this process, but no one knew whether the degree of vacuum impregnation was critical, or even adequate. Mr Fleming asked Burch to study the problem and he set about it with characteristic thoroughness; he tried to impregnate a piece of pressboard in the best vacuum he knew, an x-ray vacuum. He failed. He found his goal was unobtainable. After the dissolved air had been extracted the oil began to distil; first came the lighter fractions almost as clear as water, then the heavier fractions, but then the oil began to crack. He devised a still to separate the fractions without molecular disintegration, a still in which all molecules leaving the liquid surface reached a cold condensing surface spaced at less than one mean-free-path from the liquid. The separate fractions were carefully collected and from a knowledge of the rate of distillation the vapour pressure of each could be calculated. One of the fractions was an oil of low viscosity having a vapour pressure of less than 10^{-6} mm and hence, he reasoned, this might be used as the operating fluid in a Langmuir diffusion pump instead of mercury, and no liquid air would be needed. One fraction resembled vaseline but had an unmeasurable vapour pressure so might be used to seal joints which fitted tolerably well. Burch had been a Greek scholar—I think he entered Caius with a classical scholarship—and he christened his low vapour pressure products 'Apiezon'. Excellent patents were taken out which, years later, permeated the pharmaceutical field and were worth, I was told, almost a million pounds. Why Burch did not get a Nobel Prize for this I do not know but the whole world benefitted rapidly from his work. He asked me to test the oil so I cleaned out the mercury from a Langmuir pump and used it to pump one of my discharge tubes perfectly satisfactorily. My colleague Philip Moon set this pump up with a pressure gauge and, without a liquid-air trap, measured the lowest pressure attainable and sent Burch the results. Burch was delighted. Later, Burch designed pumps specially for his oils, and I and Cockcroft and Walton used them long before they came on the market. He was elected an FRS in 1944 primarily for this work.

In his first Presidential Address to the Royal Society on St Andrew's Day 1927, Rutherford dwelt 'on investigations carried out in recent years to produce high voltages for general scientific purposes'. He referred 'to the use of cascade transformers going up to 2 MV (RMS); while no doubt the development of such voltages serves a useful technical purpose, from the purely scientific point of view interest is mainly centred on the application of these high potentials to vacuum tubes in order to obtain a copious supply of high-speed electrons and high-speed atoms'. He spoke of Dr Coolidge's work in Schenectady—which had inspired me—and made the point that whilst the individual energy of an electron in a tube might be far below that of a beta-particle, a current of one milliampere is equivalent to beta-rays from 100 kg of radium, and concluded 'It has long been my ambition to have available for study a copious supply of atoms and electrons which have an individual energy far transcending that of the alpha- and beta-particles from radioactive bodies. I am hopeful that I may yet have my wish fulfilled, but it is obvious that many experimental difficulties will have to be surmounted before this can be realised, even on a laboratory scale'. He got his copious supply of atoms far earlier than he ever expected; they did not transcend the energies of alpha- and beta-particles but they were highly effective for his purpose which was to explore the nucleus.

Early in 1928 Rutherford asked me if it would be possible to put a complete voltage generator and vacuum tube under oil, perhaps under pressure too, in our room, in order to get far higher voltages on the discharge tube than I was then getting—something over 450 kV, albeit with many breakdowns. Brian Goodlet gave a lot of thought to this; I have his sketches as I write, they show a great Tesla transformer for 2 MV in a steel pressure vessel. He asked my opinion about the strength of the glass discharge tubes I was making, and Mr McKerrow, somewhat of a porcelain expert, suggested a porcelain tube—one was actually made—instead of glass. Now Goodlet had had many discussions with Rutherford who thought very highly of him; whenever Goodlet came to see me the Professor would invite us to his usual Sunday afternoon tea party. Goodlet had shrewd powers of observation. There is a delightful sentence in one of his letters (3 March 1928) about the generator under oil which I will quote:

> Working on these lines and assuming an eighty per cent improvement in the oil, I think that a 2000 kV outfit could be put in a tank about 6' in diameter

and 9' in height. I suppose the Professor will want these dimensions halving; that man ought to be made to look at things through the wrong end of a telescope since everything is too big for him.

Later that month Mr McKerrow discussed the project with Tizard of the DSIR to find out if the Department would support the scheme financially as it was too big for Fleming to give to the Cavendish and finance out of his own departmental budget. Although the project occupied much of Goodlet's and my time during the summer of 1928 other concepts arose which finally put this in abeyance.

During the early summer of 1928 Goodlet consulted John [Cockcroft] and me about his sealed-off cathode-ray oscillograph required to study the high-voltage transients on which he wanted to work. The oscillograph had proved to be of inadequate writing speed to record details of a surge lasting for a few microseconds. In Germany, Professor Rogowski at Aachen and Dr Gabor at the Technisches Hochschule of Berlin had built, independently, oscillographs operating with 50 kV on their cathodes, with internal photography, and whilst being pumped continuously with mercury diffusion pumps. The electron beams were focused with short solenoids and extremely fast writing speeds had been achieved. Cockcroft, writing to me from the British Association in Aberdeen, suggested that we might help Goodlet by assembling a discharge tube with a hot filament and apply a focusing solenoid to the electron beam accelerated to 50 kV. The German tubes used cold cathodes. As I was going to Germany for my holiday that summer Goodlet suggested that I should call on Rogowski and Gabor to discuss their achievements, the Company would pay my travel expenses, he said. I sent Brian a long report on my observations so that F P Burch was able to start construction of a pumped cathode-ray oscillograph with a cold cathode, very similar to Rogowski's but with important additions incorporated by Gabor (who, years later, wrote to me to seek my help in leaving Hitler's clutches—and invented holography). There was no need for us to do any experimental work in the Cavendish but I was glad to be able to repay Metropolitan-Vickers in a small sort of way for the benefits I had received.

In the middle of 1928 Gamow, a Russian theoretical physicist—who I am sure I was told had escaped from Russia by rowing across part of the Black Sea†—published a paper in the *Zeitschrift für Physik* on his theory which related the range of an alpha-particle to the half-life of the parent atom, a theory based on the new quantum mechanics. The greater the range, i.e. the greater the energy of the particle, the greater the probability it had of escaping *through* the energy barrier surrounding the nucleus, its energy being insufficient to *surmount* the barrier. The greater the probability of escape, the shorter the half-life of the disintegrating nucleus; the well-known Geiger–Nuttall law. He then wrote another paper (in German) which was sent in

† That particular attempted failed: Ed.

manuscript form to Cambridge in December 1928 on the theory of artificial disintegration, a paper later published in the *Zeitschrift*.[2] In this paper he suggested that the inverse wave-mechanical concept might also hold, that the alpha-particle (the helium nucleus) might penetrate *through* the barrier into the nuclei of light elements. This, he suggested, was the reason why Rutherford in 1919 and Rutherford and Chadwick in the 1920s had succeeded in transmuting many atomic nuclei, up to aluminium in weight, with alpha-particles. As far as I know this information was not immediately discussed in the Cavendish but was the subject under discussion at the meeting of the Kapitza Club on 29 January 1929 and at the laboratory physical society on the following Thursday. I well recall—as I have recounted in the Royal Society Rutherford Lecture[3]—returning from the Gamow colloquium to the room in which John [Cockcroft] and Ernest Walton and I worked, and he and I stood round John as he put into the Gamow formula 1 microamp of protons, which seemed a suitable figure, accelerated to, let us say, 300 kV, well below the voltage my tubes were withstanding, and letting the protons bombard a target of lithium (or boron, I cannot remember exactly). Making generous allowances for the loss of protons as the beam emerged from a Lenard window the number penetrating the energy barrier seemed sufficient to give an observable number of disintegrations. Sir John died in 1967 and when Mark Oliphant was writing the Royal Society Biographical Memoir of Cockcroft's life he found among the Rutherford papers in the University Library a memorandum written by John late in 1928 to Rutherford, *after* he had read or had heard of the Gamow manuscript, in which he put the figures I have quoted above but mentioned boron. I had never heard of this Gamow paper and did not know that John had sent a memorandum to Rutherford; the surprising thing is that, as John was alive when the Rutherford lecture was published he never corrected my account or added to it in any way. Mark did not know of the existence of it till he was writing the Memoir in 1968 and Ernest Walton had never seen a copy till I sent him one.

The upshot was that Rutherford gave the 'go ahead' and John started to assemble apparatus. He decided not to use a Tesla transformer for two reasons; in the negative half-cycle of the supply there might arise autoelectronic emission effects which might confuse any measurements, and secondly, with a damped oscillatory wave the real time available for accelerating particles at the highest crest of the waves was only about one thousandth part of the total time, so he elected for a DC supply. Even with a voltage-doubling circuit this still necessitated an AC voltage higher by far than any available in the Cavendish so once more he turned to Brian Goodlet. The problem was to get a 350 kV (crest voltage) transformer which could be got into our room; any standard Metropolitan-Vickers testing transformer of this voltage would not even be able to get through the arch from Free School Lane let alone the poky entrance into our room, even with the doors removed. Goodlet came to the rescue. He started a design of transformer specially for the Cavendish but all in good time it was adopted for a wide range of transformers which filled the need of the x-ray market. It consisted of making coils by a new technique; fine copper wire was wound spirally outwards on a flat, wax-impregnated paper sheet impressing the wire into the surface covering of wax. Many layers were wound on top of one another till a thick pancake of about 2 cm depth was made, and then machined and wrapped with insulating tape made oil-proof by varnish. Many coils made up the total required secondary winding. These were first made in the workshop of the High-voltage Laboratory and later went into production. To keep the total height of the transformer lower than the ceiling Goodlet devised a special bushing set at an angle to the core. The transformer had to be shipped to the laboratory in parts and assembled there, there were no facilities such as overhead cranes to give any help in putting the unit in place, and the transformer behaved perfectly from the word go. Walton stopped his work on indirect methods of electron acceleration, and joined John. I could not do the same as my PhD thesis had to be on acceleration of electrons but I did design the rectifiers for them (British Patent 366 561) based on my discharge tube designs. Burch provided the oil diffusion pumps and Oliphant's design for a positive ion source was, I believe, used. All worked well and the apparatus is described in the *Proceedings of the Royal Society* for 1930.[4] It can be said without any doubt that this experiment was designed to test a theory; without Gamow's second paper the work of aiming at producing disintegrations at the paltry voltage of 300 kV or thereabouts would probably never have been started. It can also be said, I think, that the Professor's words in his Presidential Address to the Royal Society alerted scientists in other countries to think about artificial disintegration, and Cockcroft and Walton only just 'made it' in time. Others were building apparatus and were ready to repeat the results achieved in Cambridge almost immediately afterwards, so Metropolitan-Vickers can correctly claim some credit for the final result.

It was during 1928 that the Vickers Company sold its shares in the Metropolitan-Vickers Company to the General Electric Company of Schenectady, the company already owning the whole of the shares of the British Thompson-Houston Company of Rugby. So the two deadly rivals were united financially though never spiritually; they remained commercial rivals till they ceased to exist at all. The new Chairman, Sir Felix Pole, until then General Manager of the Great Western Railway, came with Mr Fleming late in 1929 to meet Rutherford and he confirmed his intention of supporting Fleming in helping the Cavendish whenever possible. He was a kind and generous man and I recall he even invited us to dine with him in Cambridge. I think it was at this visit that Rutherford was invited to Manchester. Metropolitan-Vickers was building an extension to the High-voltage Laboratory converting a half million volt transformer to a one million volt unit, erecting a million volt impulse generator, and a million volt Tesla transformer, and decided to have an opening ceremony. This was to be held on February 28 1930 and Rutherford was to be the chief guest. A huge gathering of some fifty pure and applied scientists was invited at the Company's expense to Trafford Park, over ten came from Cambridge alone, all members of the Cavendish. The photograph taken in the laboratory is shown in figure 3.6.1; it includes four Nobel Laureates and three others who later won Nobel Prizes.

What is very interesting is an analysis of Rutherford's speech. First, the Chairman, Sir Philip Nash said that Metropolitan-Vickers had always recognised that the closest co-operation must exist between the activities of science and engineering, indeed the whole industry grew out of scientific research. Sir Ernest recognised that the installation was of great value to the industry but, he said, it afforded a unique opportunity for researches on fundamental problems. He was most specially interested in the application of very high potentials to discharge tubes to obtain copious supplies of electrons and atoms of high energy; he mentioned potentials of 3 and 5 million volts and he hoped such voltages may be produced soon.

I must point out, however, [said Rutherford] that the ordinary university laboratory with its exiguous finance cannot hope to erect a cathedral-like structure such as we see today to house the high potential installation. What we require is an apparatus to give us a potential of the order of 10 million volts which can be safely accommodated in a reasonably-sized room and operated by a few kilowatts of power. We require, too, an exhausted tube capable of withstanding this voltage and I recommend this interesting problem to the attention of my technical friends.

Now some eighteen months earlier Cockcroft and Walton had started to generate, not 10 million volts, but less than 350 kV with which they hoped— or rather Gamow prophesied—that transmutation might be achieved. Rutherford never mentioned this as a possibility and I think he did not do so because he did not firmly believe in the theory in spite of having let the experimenters build a generator for a mere 350 kV.

I have mentioned the firm friendship which had developed between the Professor and Goodlet; indeed, Goodlet loved Cambridge and all it signified. His education had been shattered by the Russian revolution, he had come to England with a League of Nations passport, stateless, and had found employment with British Westinghouse in their Sheffield works, attending odd courses but had never had a chance of studying for a degree. His mathematics was of an extremely high order, he was nicknamed 'Proteus' after the great American electrical engineer, Charles Proteus Steinmetz, the authority on electrical transients, and suddenly he wanted to start all over again, even though at that very moment he had almost completed the finest high-voltage laboratory in Britain. He discussed with Mr McKerrow and

Figure 3.6.1 Opening ceremony of the Metropolitan-Vickers High-voltage Laboratory extension, 28 February 1930. Cavendish members present include Rutherford, C T R Wilson, Aston, Fowler, Walton, Blackett, Cockcroft and Kapitza. Photograph courtesy Professor T E Allibone.

Left to right: *Front row:* C T R Wilson, C E Lloyd, G E Bailey, Sir R Glazebrook, Sir P A M Nash, Sir Ernest Rutherford, A P M Fleming, Lord Verulam, J S Peck, W L Bragg, B L Goodlet. *Second row:* C W Marshall, S R Milner, L St L Pendred, F W Aston, F E Smith, Miles Walker, E N da C Andrade, V E Pullin, H H Johnson, S Donkin, K G Maxwell. *Third row:* B G Churcher, J C Whitmoyer, J B Hansell, A G Ellis. *Fourth row:* W C Clinton, C G Goodwin, C G Bannister, H Warren, E B Moullin, D H Hartree, E W Marchant, Father O'Connor, R H Fowler, K Baumann, J Billington. *Fifth row:* W O Fenwick, E T S Walton, A E Barclay, G H Lepper, E B Wedmore, C H Desch, T E Allibone, P M S Blackett. *Back row:* E H Rayner, L B Atkinson, A E Tanner, J D Cockcroft, P Dunsheath, R W Lunt, L J Luffingham, W L Randall, E Mallett, H L Guy, C L Fortescue, Sir R Gregory, P L Kapitza.

with Cockcroft his idea of taking a Cambridge Tripos, and then he discussed it with Rutherford. Rutherford's advice was that the mind of an older man was not nimble enough to tackle the Mathematical Tripos. But finally he decided to chance his arm and again Fleming—generous as always—told him he could have two years leave of absence on half pay if he would continue to visit the Works from time to time. So early in 1930 the question of his successor occupied Fleming's mind and to my very great pleasure the post was offered to me in March. Mr Fleming wanted me to be with Goodlet as long as possible before he handed over and because I was no specialist in many matters which needed urgent attention, I had to learn the engineering requirements of the Company in a short time. I therefore tried to wind up my work in the Cavendish as rapidly as possible and to spend every available day in the Works with Goodlet, who finally entered Cambridge for the Michaelmas Term.

The room in which we had worked belonged to the Physical Chemistry Department and it had to be handed back later that year and Cockcroft and Walton found another higher room not far away. In moving they took the opportunity of designing a generator for twice the voltage. I never asked them whether they had lost faith in Gamow—why double the voltage why not halve it? (We shall see in a moment what happened when Gamow was vindicated.) Cockcroft invented, or thought he had invented, an extension to the Greinacher voltage-doubling circuit, a voltage-multiplying circuit which could give 2, 4, 6 or any multiple of the transformer rectified voltage. He and Walton made up a model and verified the calculated voltage regulation and ripple with different loads and then made a four-stage rectifier to give 600 kV and it was with this set-up that the great experiment of 1932 was performed. The Metropolitan-Vickers patent agent patented this for Cockcroft at the Company's expense, the Company reserving the right to use the circuit freely for its own installations, but agreeing to pay a royalty on sales of generators employing the circuit. Whether John was still receiving an emolument or, as sometimes called, a retainer or consultancy I do not know; the financial dealings were never my concern and I never enquired in later years but he was certainly regarded as one of the family and one of Mr Fleming's 'boys'.

Even before I left Cambridge Professor Barclay, the radiologist, had asked me if I could make an x-ray tube, continuously evacuated, to operate at very high voltages, say up to 1 million, as the St Bartholomew's Hospital radiotherapy unit was looking for increased depth-doses for cancer therapy. A flourishing business in such tubes and pumped rectifiers developed for 250 and 500 kV and finally for 1 MV and for this the voltage-multiplying circuit was needed. I think it was Goodlet who found that either Greinacher or Schenkel had invented this circuit in 1919 and the Cockcroft version differed in such a small amount that the patent agent considered that it had been anticipated. Nevertheless, the Company had made the patent agreement with him and we thought it was only fair, in spirit if not in law, to honour the agreement. I continued, as others did, to call it a Cockcroft–Walton circuit though I knew that we really owed the title to the German inventor. The Philips Company knew this and never paid any royalty to the British scientists; there was no need to do so.

In the summer of 1931 the Society for the Promotion of Cultural Relations between the USSR and Britain arranged a tour for scientists to the Soviet Union. It must have proved popular because ultimately two tours were arranged, primarily, I think, by J G Crowther, Science Correspondent for the *Manchester Guardian*, a friend with whom I had had many arguments about the evils of communism, though we never agreed. John wanted to go and recommended that I should go too, though the Metropolitan-Vickers Company had had a long-term agreement to teach turbine manufacture in Leningrad. It was finally decided that I should accompany him but because of conflicting appointments I went on the first tour and he on the second, the Company paying his costs. All I need say here is that we saw many of the achievements in science, some very impressive, but neither of us was any more enamoured of communism than we had been before, unlike so many of our Cambridge fellow scientists. We both met the Metropolitan-Vickers staff in Moscow and this helped us to keep a sense of proportion not kept by so many of the Cambridge post-graduates, many of whom never even went to the country.

In the autumn of 1931 Rutherford was, I think, *en route* to Ireland and Mr McKerrow invited him to spend the day in the Metropolitan-Vickers research department. There is a photograph extant showing Mr Fleming and Mr McKerrow, with some of the staff, C R Burch (FRS 1944), and R W Bailey, the head of the mechanical and metallurgical sections of the Department, who in due time became the sixth of Fleming's trainees and staff to be elected FRS (in 1949). John had been the first, in 1936, and years later three more members of staff were elected. I do not know whether this would qualify for the *Guinness Book of Records*, nine from one industrial laboratory, but I do know that all were greatly indebted—and none more so than John—to Fleming for the wonderful start in life he gave to us all. During his visit I showed Rutherford, by now Lord Rutherford, the experiment I had just done for Basil Schonland on the branching of lightning. To do this I seated him in an enclosure of black cloth and there he watched million-volt sparks with his usual exuberance but I was a little concerned lest his gesticulations might bring his arms just too near to a million volts; he thoroughly enjoyed himself, for this was the first time he had encountered rather terrifying sparks at very close quarters.

As far as I can recall, the discovery of the neutron, announced in February 1932, made no impact on us in Trafford Park. We shared in the excitement, of course, but we probably saw no direct connection between this and our work or our industry. Later that spring the story was different. Walton had told me how they were looking at the tracks of protons brought out of the discharge tube into the air, extending some 5 cm from the window of the tube but he had no indication of the impending discovery. When it came, the world was overjoyed and, of course, especially pleased were their many friends in the Company; it provided a ray of sunshine through our gloom, for we had just voluntarily taken a ten per cent salary cut to help the Company weather the Depression (how different is the attitude in 1983). Mr Fleming was interviewed by the Press and paid the highest compliment to the two scientists; he spoke of the help his Department had given them, but as far as I recalled at the time, he spoke without exaggeration though the Press used words which I am sure he had not employed.

In recent years I have written the history of the Royal Society Dining Club and had occasion to examine all the Club's 250 years of archival material. On April 28, a week after the news had been reported at the Royal Society, Rutherford invited Cockcroft to the dinner of the Club; R H Fowler invited Walton, Aston invited Chadwick, McLennan invited Dirac, Jeans invited Mott and Robertson invited C D Ellis and a total of 24 sat down. Now at the first dinner of the Session on 5 November 1931, General Smuts, President of the British Association had been invited as the Guest of the Club, a rather special honour, and the Treasurer, whose prime duty at the dinner is to keep a record of any interesting items of science mentioned in the after-dinner discussions, recorded fairly extensively some remarks by Lord Rutherford; 'Lord Rutherford drew attention to the remarkable period of advance in physics, which had seen the advent of three new theories of mechanics, each of which was an advance on the other. But the theorist had however come to a halt before the problems of the nucleus and a still more fundamental theory was awaited'. (Note he never mentioned the exciting predictions of Gamow, that particles may soon be found entering the nucleus, once the experimental conditions are right.) Within four months of this comment Rutherford's son-in-law, Professor R H Fowler speaking at the Club described a revolution, not in theoretical physics, but in experimental physics. The entry for 3 March reads 'Experiments in Cambridge have led to the identification of a new subatomic entity consisting of one proton and one electron to which the name "neutron" has been given'. This was followed two months later by the wonderful news of atomic disintegration, another experimental achievement, albeit owing all its inspiration to Gamow's theory. At dinner, Rutherford referred 'to the important part which is being played by the younger men in connection with the elucidation of the structure of the atom'; he was very proud of his 'boys', as we were all called by him, and Cockcroft and Walton were very specially treated by him in being invited to the Club that night.

Our reaction to the great news [the Cockcroft–Walton experiment] was to re-examine what we should do; was the time ripe to start any disintegration experiments? We decided not to do so. After all, we were an industrial

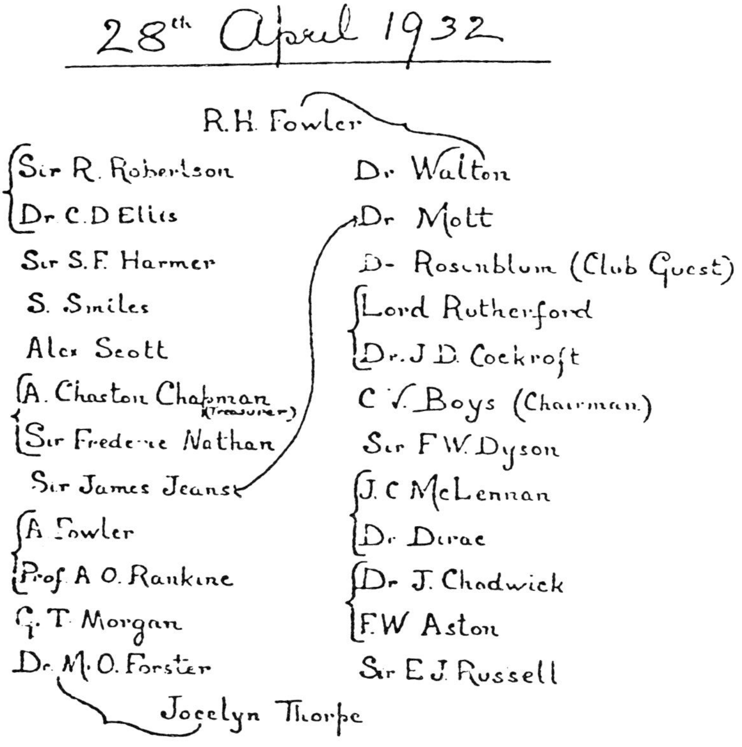

Figure 3.6.2 Seating arrangement for the Royal Society Dining Club, 28 April 1932, following the discovery of the neutron. Courtesy Professor T E Allibone.

laboratory with urgent commitments and though we sometimes did some basic science arising out of our normal work, we did not initiate basic work far removed from the Company's main interests. Moreover we were suffering severely from the Depression and had had to lose members of staff; the times seemed out of joint. The inefficiency of the nuclear reaction was extremely great, a few kW to produce a few microwatts. It was not until the chain reaction had been discovered in 1939 that this situation changed.

The greatest effect was on Rutherford himself; Gamow now was fully vindicated, the 'old man' at last believed that low energy particles could penetrate the nucleus and he decided that, whilst Cockcroft and Walton struggled with higher and higher voltages, he would cash in on *lower* voltages. He asked Cockcroft to talk to me about the urgent supply of one or two transformers of modest voltage but then decided he would write to me direct, to speed matters up:

> In our experiments on artificial disintegration I would like to have, for experiments, one of these transformers for 75 000 volts. I think you know the requirements of the laboratory. We want the weight reduced to a minimum, and we do not expect to run the transformer continuously. I imagine an output of 100 mA would suffice. I do not know to what stage you have got in the design or manufacture of these transformers. Please let me know when you think you could deliver one as I would like to try it out as soon as possible. I think you know I want to avoid bulky units so that we are able to transfer them from one room to another. I should be glad if you would let me have your views on the question of time of delivery and probable price. I am leaving for Wales today, drop me a note to my cottage.

I replied the following day and two days later came:

> I note there is not much difference in weight between transformers for 75 kV and 100 kV, the latter would give 200 kV with doubling. Have you a 100 kV transformer ready to send off; what is the price? Let me know.

(It was not correct for me to send quotations which committed the Company legally, I could only hint a price and ask the correct department to send a quotation.) Back came his third letter a few days later:

> I would like to have at <u>once</u> [his underlining] the 100 000 volt transformer to give 200 000 volts with doubling and I note the price is £85, so send it to the Cavendish as soon as you can, so we can arrange for the experiment immediately. I shall be obliged if you can hurry the delivery as our work is held up till we get it.

Both letters from Wales were handwritten, followed by a nice letter which I have lost thanking me for the great speed of delivery. Later that year I called on him in Cambridge and he twitted me about the price. I did an instant calculation and snapped back 'But Prof, that is only one farthing per volt and you pay three pence per volt for a flash lamp battery'; he accepted that in good grace and invited me to dine in Trinity. The world knows how he, with Mark Oliphant as junior collaborator, produced the D–D reaction at voltages of around 100 kV and he even demonstrated that experiment at a Friday Evening Discourse at the Royal Institution. My small, very small, transformer can be seen at the extreme right of the apparatus photographed as he was lecturing (figure 3.6.3), we had indeed made it portable as he had asked.

I cannot recall many close connections with the Cavendish in later years. John and, later in the same month, Walton stayed with me in 1933 and both told the story of how they had been virtually kicked into doing the lithium bombardment experiments by the Professor instead of 'messing about'

Figure 3.6.3 Rutherford demonstrating deuterium fusion at the Royal Institution, 1934. The Metropolitan-Vickers transformer is to the extreme right of the apparatus. Reproduced by kind permission of Sir Mark Oliphant from his book *Rutherford: Recollections of the Cambridge Days* (Amsterdam: Elsevier, 1972)

measuring the range of the protons in air. Many pieces of apparatus were sold to the Cavendish including a magnet of special design, parts of the 37 inch cyclotron, and replacement apparatus after Kapitza had been forcibly detained in Russia in 1934 (note that he had always kept his Russian citizenship and passport, so he could not complain if they kept him) and the original machine had been sent out to him. Radar came as the war clouds grew darker, the 40 Chain Transmitter stations were to be built around the coasts using very large transmitting valves oscillating at high frequency. No suitable valves were available but Burch had made a 500 kW valve for the Post Office which was evacuated continuously, the design being based on his techniques for grinding large flat matching surfaces to high degrees of precision and then sealing the joints with Apiezon grease. Valves like this could do the radar job. I happened to be in the Cavendish when the Ministry contract was being negotiated with Metropolitan-Vickers and Rutherford, who, I think, was on the defence committee concerned, asked me a lot about the reliability of operation of large units pumped continuously and I was able to assure him of the completely satisfactory operation, from this point of view, of many 250 kV x-ray tubes operating eight hours a day over many years. He was glad to hear of my confidence in the pumped systems and spoke of his great admiration for Burch and of the debt the Cavendish owed him for his work. As I have already written, the Battle of Britain would probably not have been won without his valves.

There are just two more connections between the Metropolitan-Vickers Company and Cockcroft, Rutherford and the Cavendish that I might recall. In 1934 Fermi had discovered that when elements are made radioactive by neutron bombardment the amount of radioactivity is increased enormously if a layer of paraffin is interposed between the source and the target. I think it was the Rector of the University of Rome who at once appreciated the probable commercial importance of this discovery. At any rate, it was at his insistence that a patent was applied for on the next day, 26 October (Italian Patent 324458) and in due course in other countries. The agent for the Italian patent holders, one Giannini, got in touch with our Chairman, Sir Felix Pole, who notified Fleming and I was sent to London to meet Giannini. I had been to Cambridge to discuss this with Cockcroft—with a D–D source of neutrons of high energy the use of a neutron velocity moderator might produce a gamma-ray source of high intensity for deep x-ray therapy which might be competitive with a radium therapy 'bomb', radium then costing £37 000 per gram. Cockcroft did the calculation with me, but I have alas lost it. We worked out the cost of an accelerator tube for various voltages, and he calculated the probable neutron and thus gamma-ray yield, but with the most optimistic figures we could not get the overall cost nearly low enough for such an apparatus to rival radium gamma-rays. John came with me to meet Giannini and then I reported back to Fleming. We thought that in this rapidly moving field of science new ideas might arise which would make the patent valueless so Fleming suggested that the Company should buy an 'option' on the patent for a three-year period, paying the University of Rome £2000 a year; John did not disagree with this view but Giannini turned it down and we let the matter drop. The patent was, of course, the basis of the moderator in a nuclear power reactor and the US Government paid Fermi's widow an 'ex-gratia' sum for the reactors at Hanford, Washington. The patent lasted only sixteen years during which time the Atomic Energy Authority had built a reactor in Harwell but I had found that there was an error in the British patent which, our patent agent said, invalidated the patent and I had to recommend (I am not sure whether this was made direct to Cockcroft or to our agent) that no payment need be made.

The other connection concerned the one million volt x-ray tube at St Bartholomew's Hospital, the DC generators of which used the Cockcroft patent on which a royalty was therefore paid. I suggested that Rutherford be asked to perform the opening ceremony of the Mozelle Sassoon Laboratory; he came on 10 December 1936 direct from the House of Lords where he had listened to the Abdication speech. When I read the paper on this work to the IEE I suggested that Cockcroft be invited to open the discussion; which he did, for there could have been no more appropriate scientist to comment on the work.

Less than a year after this opening ceremony Rutherford was dead. After the interment of his ashes in Westminster Abbey I returned to our London office with Sir Felix Pole and Fleming who spoke of his long connection with Rutherford, starting in 1915 and extending through his long association with the DSIR of which Rutherford had been Chairman for many years, and he specially referred to his admiration of Rutherford's great modesty in the midst of his achievements. It was a sad day. Perhaps after all these years I may be permitted to draw back, slightly, a curtain covering one aspect of Fleming's life which must have given him pain. In 1939 Professor Miles Walker submitted Fleming's name for election to the Fellowship of the Royal Society and almost at the same time the name of C C Paterson, head of the GEC research laboratory was also proposed and, of the two, Paterson was ultimately elected. Both had, without doubt, rendered conspicuous service to the cause of science, but Paterson, had, as a young man, done 16 years of research in the NPL whereas Fleming had spent the whole of his life in industry, running part of a large factory from which he developed the research and education activities I have described. Of the two of them—and I knew Paterson well—I should have given Fleming the pride of place and I feel certain that, had Rutherford been alive, he would have done likewise. Fleming never mentioned this, of course, but as his 'boys'—and there were nine of them†—were elected one by one, he must have felt sad whilst

† There might have been ten; Brian Goodlet's name was proposed in 1961. Alas he died in October of that year.

rejoicing that they had trodden a road on which he would dearly have liked to have walked.

The story of the Company's connection with the Cavendish is a happy one. There was no ulterior motive in the Company's interest in the development of science there. This interest began simply because Fleming and Rutherford had worked in harness during the First World War, and because Miles Walker had advised John Cockcroft, already one of Fleming's apprentices, to take the Cambridge Tripos and Fleming being a kind and generous man keen on the training of school leavers and graduates had offered a mere £50 to help his further education. Fleming's own laboratory in 1922 was small and contained only a few disciplines, so it was natural to turn to Bragg to guide A J Bradley, to von Hevesy and to Desch to guide me and Charles Sykes, to Rutherford for guidance in atomic matters and so on and so on. It was *not* done for commercial gain; he gave apparatus to many universities and sold apparatus on generous terms to friends in many universities, and was at all times supported by his Chairman, Sir Felix Pole. The Cavendish gained; but we gained too, goodwill, friendship and a freedom to go to the Cavendish for inspiration and guidance.

Notes

1   Kapitza P 1927 *Proc. R. Soc.* A **115** 658
2   Gamow G 1929 *Z. Phys.* **5** 214
3   Allibone T E 1964 Rutherford Lecture *Proc. R. Soc.* A **282** 447
4   Cockcroft J D and Walton E T S 1930 *Proc. R. Soc.* A **129** 477

3.7 Theoretical Physics in Cambridge in the late 1920s and early 1930s

Alan Wilson

In the early 1920s Cambridge was in a state of flux pending the outcome of the Royal Commission on Oxford and Cambridge, which resulted in the formal teaching of all subjects being transferred from the colleges to the University in 1926. Up to that time, while Natural Science subjects had been organised on a University basis, all arts subjects, including mathematics, had been college-based. In the teaching of mathematics the colleges were divided into three groups and in general it was not until one's third year, when reading for the Mathematical Tripos Part II, Schedule B (now called Part III), that one went to lectures in colleges outside one's own group.

This separation of subjects into college-based and University-based groups meant that no mathematician was allowed to attend any lectures on physics (and *a fortiori* not permitted to witness demonstrations or perform experiments) and even the phrase 'theoretical physics' was never used, it being considered a minor part of 'applied mathematics'.

I graduated in 1926, the year of the General Strike, and it was only because the University Appointments Board had failed to find me a job that I decided to stay on and do research. E Cunningham of St John's College offered to be 'supervisor' but I eventually managed to attach myself to R H Fowler. In June he gave me a list of papers to read, and in October he decided that I should tackle the 'two-centre problem' (the energy levels of the hydrogen molecular ion) which he found to be too difficult to solve himself.

Fowler was a large and vigorous, but somewhat undisciplined, man. As College Lecturer at Trinity he had a room in Neville's Court but he lived at Trumpington, where he did most of his work. If you wanted to see him, you called at Trinity on the off-chance of finding him there. If he wanted to see you he sent a handwritten note to your rooms by the Trinity porters. But he spent a lot of time in London, Germany, and even further afield. So one often had to struggle on one's own when one had reached an impasse. I remember once receiving a note from him posted in Trumpington asking me to call on him in Trinity. When I got there his room was unoccupied, but there was a note on the table saying 'Back in fifteen minutes'. When I had waited an hour, I made my way to the Porters' Lodge to be greeted by the information that 'Mr Fowler left for America yesterday, and it is not known when he will be back'.

In the academic year 1926–7 there were only three research students working on quantum mechanics, J A Gaunt and W H McCrea of Trinity, and myself. My contacts with Gaunt and McCrea were minimal, and they were confined to weekly seminars, held in the Cavendish on Wednesday afternoons, which were devoted primarily to nuclear physics but which also attempted to present the salient features of a few of the papers appearing in the German periodicals following the discovery of wave mechanics by Schrödinger. Dirac was spending a sabbatical year in Germany, and Fowler was the only person qualified to distil the essence of this spate of discoveries. But he was heavily overloaded. The discussions were therefore perfunctory, except, of course, on experimental nuclear physics.

By the early 1930s the situation had improved considerably. From 1929 until 1940 I gave a course of lectures in the Michaelmas term, entitled 'Quantum Theory' and based upon applications of the Schrödinger equation. This was supplemented by a course in the Lent term by Dirac (except when he took sabbatical leave) covering his book *The Principles of Quantum Mechanics*, published in 1930. Furthermore, Fowler was appointed Plummer Professor of Mathematical Physics in 1932, and he made his headquarters in the Cavendish, where he was more accessible than in Trinity or at Cromwell House, Trumpington. More or less simultaneously Dirac was appointed Lucasian Professor of Mathematics in succession to Joseph Larmor who had been inactive for many years.

The passage of time had resulted in quantum mechanics being adequately represented in the Faculty of Mathematics, there being two professors (three if one included Lennard-Jones's Chair of Theoretical Chemistry), two lecturers, an occasional assistant lecturer, some research fellows, and a reasonable number of students. In the period up to the outbreak of war, however, only four research students wished to work on the subject which interested me most, solid state physics. These were J W Harding, K Mitchell, R E B Makinson and G P Dube.

It was not until the late 1930s that colloquia were held which everyone could attend. For a considerable time, the Faculty of Mathematics had had no premises of its own. But eventually we were allocated the Arts School on the New Museums site, adjacent to the Cavendish Laboratory, the Chemical Laboratories, and the Cambridge Philosophical Society's Library. It then became possible for the first time to hold regular colloquia which ensured that all main advances in theoretical physics (and in other branches of pure and applied mathematics) could be discussed in depth.

Part 4

Change and Continuity

4.1   Introduction

The year 1932 marked the climax of the golden age of the Cavendish Laboratory, but just as the discoveries of that year were founded on the achievements of the preceding years so they in their turn acted as the foundations for further successes in both Cambridge and elsewhere. In this part we look at the continuity of success in the form of two further achievements of the Cavendish, and at some of the changes that marked the end of the golden age.

Following the achievements of Cockcroft and Walton, Mark Oliphant was called in by Rutherford to collaborate with him (or rather to work for him!) on the setting-up of a second accelerator working at slightly lower energies than the Cockcroft–Walton machine, but with greater flexibility. Using the transformer supplied by Metropolitan-Vickers, electronic recording apparatus produced by Lewis and Wynn-Williams, and an accelerating tube of his own design, Oliphant quickly set up the second accelerator, and in 1933 he and Rutherford used it to investigate the disintegration of light atoms by deuterons. At the same time as Rutherford had predicted the existence of the neutron he had also predicted that of the 'diplon', or heavy hydrogen nucleus of atomic mass 2. In 1932 in America, Gilbert Lewis succeeded in isolating this isotope of hydrogen, which he called deuterium (the deuteron being its nucleus),[1] and early the following year he presented some of the new substance, still only available in very small quantities, to Rutherford. Oliphant's new accelerator tube (figure 4.1.1) was curved so that in the presence of a variable magnetic field protons of any given energy range could be selected from those accelerated, and this also made possible the selection of deuterons. Working under Rutherford's supervision, Oliphant first repeated the Cockcroft–Walton disintegration experiments for deuterons in place of protons.[2] Then an Austrian chemist, Paul Harteck, who was working in the Laboratory, was called upon to prepare some samples of heavy hydrogen compounds that could be used as targets for the accelerator. The outcome of this, as described below by Oliphant, was the fusion of pairs of deuterons, each producing either helium-3 nuclei and a neutron or tritium (hydrogen-3) nuclei and a proton, in either case with a sizable release of energy.[3] The small probability of the reaction meant that while each individual fusion event was energy producing the experiment as a whole was a massive energy consumer, and it was in this context that Rutherford dismissed atomic energy as 'moonshine'.[4] But the demonstration of deuterium fusion turned out to be of enormous importance, both for the understanding of natural fusion processes in the stars and for the production of artificial fusion energy.

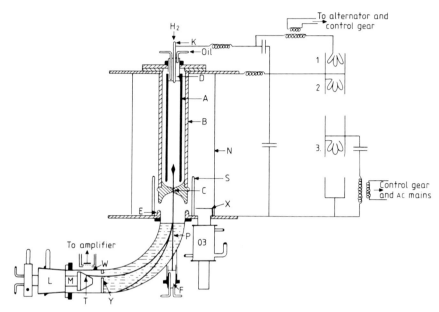

Figure 4.1.1 Oliphant's accelerator tube of 1933 (that used for deuteron fusion in 1934 differed only in being mounted horizontally, and in the power supply). The core of the apparatus is a version of a canal ray tube with A and C as the electrodes. The beam of protons and other positive ions generated in the canal ray tube passed through the perforated cathode C towards F. An accelerating potential was applied between C and E and the accelerated beam was then bent round in a magnetic field towards the target T. By varying the magnetic field Oliphant could ensure that only ions of a given energy reached the target, others being deflected more or less. Redrawn from *Proc. R. Soc.* A **141** 261 (1933).

The deuterium fusion results were published early in 1934, and during the same year there was also another major discovery in the Cavendish, again using the new heavy hydrogen. Following up a suggestion of a young refugee research student, Maurice Goldhaber, Chadwick and Goldhaber demonstrated the photodisintegration of deuterium, or the nuclear photoelectric effect. As Goldhaber recalls below, this was not only of interest in its own right but also led to the first accurate figure for the mass of the neutron and to a speculation, unfortunately suppressed, as to the significance of slow neutrons. This was some months before Fermi stumbled upon and published details of the same phenomenon. The work of Chadwick and Goldhaber also

Figure 4.1.2 The Seventh Solvay Congress of Physics, Brussels 1933.

Standing from left to right: E Henriot, F Perrin, F Joliot, W Heisenberg, H A Kramers, E Stahel, E Fermi, E T S Walton, P A M Dirac, P Debye, N F Mott, B Cabrera, G Gamow, W Bothe, P M S Blackett, M S Rosenblum, J Errera, E Bauer, W Pauli, M Cosyns, J E Verschaffelt, E Herzen, J D Cockcroft, C D Ellis, R Peierls, Auguste Picard, E O Lawrence, L Rosenfeld. *Seated from left to right:* E Schrödinger, I Joliot-Curie, N Bohr, A Joffé, M Curie, P Langevin, O W Richardson, Lord Rutherford, Th De Donder, M de Broglie, L de Broglie, L Meitner, J Chadwick. Photograph courtesy of the Cavendish Laboratory, University of Cambridge.

led, as mentioned by Rudolf Peierls below, to the important theoretical analysis of the deuteron by Peierls and Hans Bethe, the first of many contributions both were to make to nuclear physics.[5]

Throughout the 1930s Cambridge physicists continued to make an important contribution to the development of physics, but even in 1934 they were already beginning to be eclipsed. The geography of experimental nuclear physics was beginning to be affected by Fermi's decision to move over from theory, and in the middle of the decade it was Fermi's group in Rome who made the running. On the theoretical side Dirac was going his own way, and mainstream developments focused on European and American centres. In the early 1930s American physicists were just beginning to make their presence felt in theoretical physics. By the end of the decade, reinforced by large numbers of refugees from Europe, they were quite dominant.

Indeed, while the development of physics continued to be rapid throughout the 1930s, it was to some extent overshadowed by political developments. Before Hitler had come to power in Germany that country had become the undisputed centre of world physics. But many of the most talented physicists were either Jewish or in other ways politically unacceptable to the Third Reich, and beginning in 1933 they were forced to leave Germany in horrendous numbers. Einstein, Schrödinger, Born, von Neumann, Szilard, Teller and Weyl were just a few of the big names forced to leave. Others included Goldhaber, Peierls, Bethe and Frisch. As Mussolini took a stronger line in Italy Fermi and others also felt they had to leave Europe. The total number of university physicists and mathematicians exiled ran into many hundreds,[6] and a good number of these passed through Cambridge in their search for a home and position. In Cambridge, Rutherford was already Chairman of the Advisory Council of the DSIR and was reluctant to give up any more time from physics. But he was also shocked and horrified by the developments in Germany, and in 1933 he led the British effort to do something concrete for the refugees, becoming President of the Academic Assistance Council (fig. 4.1.3). Others in Cambridge, most notably Blackett and his wife, did everything they could in the way of finding accommodation and immediate relief, no easy matter in the Depression.

Given the high concentration of talent already in Cambridge, it was inevitable that most of the refugees would move on to universities able to offer them better prospects. But other changes at the Cavendish were more permanent. In 1933 Blackett left to take up a Chair at Birkbeck College, London, and two years later Chadwick was appointed to a professorship at Liverpool. In between these two departures, Kapitza was detained in Russia. In 1937 Oliphant moved to a Chair at Birmingham and the same year Rutherford died. Within the next few years the era was finally closed by the discovery of atomic fission and the outbreak of World War Two. The golden era of nuclear physics in the Cavendish Laboratory had come to an end; but the achievements of 1932 had transformed physics and the world.

Figure 4.1.3 Rutherford appealing for funds for the Academic Assistance Council at the Royal Albert Hall in 1933. Beside him is Einstein. Reproduced by kind permission of Sir Mark Oliphant from his book *Rutherford: Recollections of the Cambridge Days* (Amsterdam: Elsevier, 1972).

Notes

1   Lewis G N and MacDonald R T 1933, *J. Am. Chem. Soc.* **55** 1297; Urey H V, Brickwedde F G and Murphy G M 1932 *Phys, Rev.* **39** 164
2   Oliphant M L and Lord Rutherford 1933 *Proc. R. Soc.* A **141** 259
3   Oliphant M L, Harteck P and Lord Rutherford 1934 *Proc. R. Soc.* A **144** 692
4   Lord Rutherford 1933 *Nature* **132** 433
5   Bethe H A and Peierls R 1935 *Proc. R. Soc.* A **148** 146; Bethe H A and Peierls R 1935 *Proc. R. Soc.* A **149** 176
6   See Beyerchen A 1977 *Scientists Under Hitler* (New Haven: Yale University Press) and Pinl M and Furtmüller L 1973 Mathematicians Under Hitler *Leo Baeck Institute Year Book* vol 18 p. 129

4.2   Working with Rutherford†

Mark Oliphant

During my first five years in the Cavendish I worked on some properties of positive ions and on the separation of isotopes by electromagnetic methods. For some time I enjoyed the cooperation of P B Moon in what proved to be interesting investigations. R M Chaudhri also worked with me. This work involved the production and acceleration of various kinds of ion.

Immediately after the first observations by Cockcroft and Walton, while my wife and I were spending the weekend with the Rutherfords at their cottage, Celyn, at Nant Gwynant, in North Wales, Rutherford told me that he wanted to exploit the new technique as fully as possible in the Cavendish. He suggested that I give up my work with positive ions and use my experience in collaborating with him to set up a second accelerating system. Naturally, I accepted eagerly the privilege of such an arrangement. We decided to aim for greater accuracy in measurement of the energies of the products of these transformations. So I designed and constructed a simple version of the Cockcroft–Walton apparatus, for a maximum energy of 200 keV, and with an improved form of canal-ray tube, giving 100 mA or more of protons. This equipment was set up in the room next to that in which Rutherford and Chadwick had done most of their work on artificial disintegration with alpha-particles. Because of the low ceiling, it was necessary to use a horizontal accelerating tube. A brick wall was erected to separate the beam end of the equipment from the high voltage area. This served to reduce greatly the intensity of x-rays in the observing region, and gave us plenty of room for setting up our measuring apparatus. At various times there worked with us A H Kempton, Miss Reinet Maasdorp, B B Kinsey and Dr P Harteck. I have already remarked that Kempton wrote witty songs for the annual Cavendish Dinner; Miss Maasdorp was from Rhodesia, a cheerful soul whose politics were of the extreme left; Kinsey in voice and manner, was a character straight from Wodehouse; while Harteck, an Austrian physical chemist who had been trained in Berlin, was a very large and

† Extract from M L E Oliphant 1972 *Rutherford: Recollections of the Cambridge Days* (Amsterdam: Elsevier)

184

handsome man with incredible physical strength. All were excellent scientists.

We used magnetic analysis of our beam to ensure that we knew both the kind of bombarding particle we were using, and its precise energy. We were fortunate to have the help of Rutherford's personal assistant, George Crowe, who was very skilled in the splitting of mica to give absorbers of accurately known air-range equivalent, and in mounting these as windows opposite our targets, or as stepped absorbers for measuring the energies of emitted particles. He also prepared alpha-particle sources of polonium and thorium C^1, the particle energies of which were known accurately, to insert in place of our targets for calibration of the equipment. His ability at all laboratory procedures, and his cheerful personality, made working with him a real pleasure.

Initially, like Cockcroft and Walton, we used a scintillation screen to observe the products of transformations. Here, the long experience of Rutherford and Crowe proved of immense value. However, this soon gave way to the electronic techniques of Wynn-Williams and Lewis. Particles were registered as sideways deflections of a spot of light reflected from the mirror of a moving ion oscillograph, of French manufacture, on a moving strip of photographic paper. This was developed and fixed, and then examined by eye to count the number of particles per unit time producing a given ionisation in the detecting chamber, i.e., a given deflection of the spot of light. The speed of the mechanical oscillograph limited recording rates to about fifty per second for complete resolution and accurate measurement of amplitude.

Rutherford's participation in the experiments was limited to discussion about what to do next, and deep interest in the results. He gave us a completely free hand in the design of experiments and running of the equipment, but he kept us on our toes all the time. Like all Cavendish equipment up to that time, ours was hastily assembled from whatever bits and pieces were available, so that it often gave trouble. Rutherford was very irritated by delays of this kind, but was singularly uninterested in finding the money to buy more reliable components. However, he was extremely pleased when things went well, giving us a triumphant feeling of something accomplished.

Usually, he came to see us twice each day, late in the morning and shortly before six o'clock in the evening. Occasionally, if something exciting was happening, he would turn up at other times. This was when something was almost bound to go wrong. With Rutherford looking over our shoulders, impatiently awaiting the outcome of an observation, the operator tended to make silly mistakes. On two occasions, Rutherford himself, whose hands tended to shake, pushed something through the mica window through which the products of transformation emerged, letting air rush into the apparatus and creating panic till we had the oil pumps cooled down and everything shut off. He was humbly apologetic, but disappeared for hours, or even days, while we cleaned up the mess and got going again.

If Rutherford appeared just at the end of a run, he insisted that the record be developed as rapidly as possible, barely allowed it to be dipped in the fixing bath, and sat at the table in the next room, dripping fixing solution upon our papers and his own clothes, as he examined the tracing. His pipe dribbled ash all over the wet and sticky photographic paper. He damaged it irreparably with the stump of a pencil from his pocket, with which he attempted to mark the soft, messy material. Searching impatiently for the interesting parts of the long record, he pulled it from the coil in Crowe's hands to fall to the dirty stone floor, often trampling on it as he got up in the end. We had then to do our best to finish fixing, washing and drying the paper strips, often damaged beyond repair. When it was possible, we concealed records from him till they had been properly processed and measured up by us, but this was impossible when he was present while the record was being taken. Once, at the end of a particularly heavy day, when the experiments had gone well, we decided to postpone development till next morning when we were fresh and we could handle the long strip in new developer and fixer without damage. Just as we were leaving, Rutherford came in. He became extremely angry when he heard what we had decided, and insisted that we develop the film at once. 'I can't understand it,' he thundered. 'Here you have exciting results and you are too damned lazy to look at them tonight'.

We did our best, but the developer was almost exhausted, and the fixing bath yellowed with use. The result was a messy record which even Rutherford could not interpret. In the end, he went off, muttering to himself that he did not know why he was blessed with such a group of incompetent colleagues. After dinner that night, he telephoned me at home: 'Er! Er! Is that you Oliphant? I'm er, er, sorry to have been so bad tempered tonight. Would you call in to see me at Newnham Cottage as you go to the Laboratory in the morning?'

Next day he was even more contrite. 'Mary says I've ruined my suit. Did you manage to salvage the record?'

He drove us mercilessly, but we loved him for it.

In 1933, G N Lewis, from Berkeley, visited the Laboratory. He presented Rutherford with about 0.5 cm^3 of almost pure heavy water, which he had concentrated electrolytically. It was sealed in three tiny glass ampoules. After much discussion, I reacted one of these with a film of potassium deposited on the walls of an evacuated glass bulb, releasing a few cubic centimetres of deuterium. Meanwhile, I asked my colleagues to try the effect of mixing hydrogen with a large excess of helium, to see whether such a mixture gave a reasonable beam of protons when used in the canal-ray tube of our apparatus. We were pleassd to find that a mixture of five parts of helium with one part of hydrogen gave precisely the same proton beam as pure hydrogen. We also arranged to collect the gas from our pumping system, and to purify it by freezing out all components other than helium and hydrogen, in a glass trap immersed in liquid nitrogen boiling under reduced pressure, to obtain as low a temperature as possible. We were surprised when the nitrogen solidified to a white crystalline substance, but the method was effective. We were thus able to use our limited supply of deuterium gas over and over again.

We found that the beam of deuterons produced a copious emission of long range, singly charged particles, which appeared to be protons, whatever target we bombarded. Even a clean steel surface produced these particles after the beam had fallen upon it for a short time. E O Lawrence had observed such particles produced by the beam from his cyclotron, and had suggested that they arose from break-up of the deuterons into protons and neutrons, in the nuclear fields of the target material. I did not believe this explanation, because the emission from a given target of steel grew with time, and concluded that deuterium was sticking to the target, so that what was observed was the product of the bombardment of deuterium with deuterons. Accordingly, Harteck prepared small quantities of compounds containing heavy hydrogen, by ionic exchange, the first being heavy ammonium chloride, and we made targets by evaporating a drop of a solution in water, placed on a steel target holder, which was water cooled. Immediately, we observed a very large emission of the long range particles, even at bombarding energies of 20 or 30 kilovolts. Rutherford was excited and encouraged us to go ahead. We made certain that the particles were indeed protons, by measuring their velocity in crossed electric and magnetic fields which we calibrated with polonium alpha-particles. I made up an ionisation chamber containing helium at high pressure. When connected with the linear amplifier, and placed near the deuterium target, many neutron recoils were observed. A shorter range group of singly charged particles equal in numbers to the number of protons, was shown to be due, almost certainly to the transformation of the nuclei of two deuterons, which fused together to give an unstable helium nucleus, into a new isotope of hydrogen of atomic mass 3, which we called tritium, and a proton.

It was impressive to experience Rutherford's enthusiasm, and the extraordinary process whereby he calculated, by approximate arithmetic, the range–energy relationship of tritium nuclei from the known range–energy curves for alpha-particles and protons. He was so impatient that he kept making slips in his argument. We showed that the momentum relationship was in accord with our assumption, and obtained a value for the atomic mass of tritium very close to that now accepted.

With great patience, instructed by Crowe, I managed to split a sheet of mica, a few square centimetres in area, which showed vivid interference colours, and had a stopping power equivalent to only 1.5 mm of air. Crowe prepared a fine-meshed grid of brass to support this mica, and succeeded in preparing a thin window which withstood atmospheric pressure. This enabled us to look for very short-range particles from our target. We found a group of particles which clearly carried a double charge and appéared to be alpha-particles, in numbers equal to the protons and tritons. This observation produced consternation among us. The equality of fluxes suggested that all three groups of charged particles originated in the same process. Rutherford produced hypothesis after hypothesis, going back to the records again and again, and doing abortive arithmetic throughout the afternoon. Finally, we gave up and went home to think about it.

I went all over the afternoon's work again, telephoned Cockcroft who had no new ideas to offer, and went to bed tired out. At 3.0 AM the telephone rang. Fearing bad news, for a call at that time is always ominous, my wife, who wakes instantly, answered it and came back to tell me that 'the Professor' wanted to speak to me. Still drugged with sleep, I heard an apologetic voice express sorrow for waking me, then excitedly say: 'I've got it. Those short-range particles are helium of mass three.'

Shocked into attention, I asked on what possible grounds could he conclude that this was so, as no possible combination of twice two could give two particles of mass three and one of mass unity. Rutherford roared: 'Reasons! Reasons! I feel it in my water!'

He then told me that he believed the helium particle of atomic mass 3 to be the companion of a neutron, produced in an alternative reaction which just happened to occur with the same probability as the reaction producing protons and tritons.

I went back to bed, but not to sleep. I called in to see Rutherford at Newnham Cottage after breakfast, and went through his approximate calculations with him. We agreed that the way to clinch the conclusion was to measure, as accurately as we could, the range of the doubly charged group of particles, and the energy of the neutrons. I went through our records from the helium pressure chamber, measuring the amplitudes of the most energetic of the helium recoils, and obtaining a maximum neutron energy of about 2 million eV, while my colleagues estimated more accurately the range of the short group. Of course, Rutherford was right. By the end of the morning we had satisfied ourselves than an alternative reaction of two deuterons produced a neutron and a helium particle of mass 3, the energy released being close to that in the other reaction. The mass of helium-3 worked out to be a little less than that of tritium.

We all shared Rutherford's excitement. We had found two new isotopes, and measured their masses, and we understood the remarkable deuterium reactions. I wrote a note describing our work that evening. This was pencilled all over by Rutherford in the morning, retyped and sent off to *Nature*. Only in the war was I to experience such a hectic few days of work, but at no other time have I felt the same sense of accomplishment, nor such comradeship as Rutherford radiated that day.

4.3   Working with Chadwick†

M Goldhaber

The Cavendish has the distinction of having freed two of the most important particles of physics from their natural state of bondage. J J Thomson freed the electron from atoms in 1897 and thus started the electronic age and 35 years later Chadwick freed the neutron and thus started what we might call the neutronic age. Now the electronic age has extended our senses and the neutronic age, we hope, will force us to come to our senses before it is too late.

I should like to tell you how it came about that I worked with Chadwick, having first wanted to be a theoretician. Let me first say a few words about Chadwick before he discovered the neutron. He was already a distinguished physicist, especially in nuclear physics: in fact he only worked in nuclear physics throughout his research life. Among his most important contributions was a discovery made when he was only 23 years old, that the beta spectrum is continuous rather than a collection of discrete lines as was then believed. He was at that time working with Geiger in Berlin. It was 1914, when the First World War started, and he was interned by the Germans for the duration of the war. During his internment he amused himself by doing experiments on radioactivity. He also interested another internee, C D Ellis in the subject. They investigated a German toothpaste which was advertised as radioactive, and for a while they thought that they had discovered a new radioactive series. The bad food at the camp ruined Chadwick's digestion, and he suffered from this for the rest of his life.

Chadwick was twenty years older than I, and he in turn was twenty years younger than Rutherford. It is hard to believe what a time lag twenty years means in physics, but if you think of it as five successive generations of research students, with each one starting a new field of physics, you can imagine what it means. I came to the Cavendish in 1933 and thus I missed the excitement at the Cavendish when the discovery of the neutron was made. At that time I was still a student at Berlin, where I had taken a course

† Talk given at the conference on the Neutron and its Applications, held in Cambridge in September 1982 to commemorate the 50th anniversary of the discovery of the neutron.

in nuclear physics given by Lise Meitner. When Chadwick's discovery became known, Lise Meitner reported on it in an extremely well-attended colloquium. She was so excited that she told us of a neutron hitting a 'brass nucleus'. But you must be forgiving because after all the 'brass nucleus' has fewer 'isotopes' namely 7, than the tin nucleus which has 10.'

Early in 1933 it became clear that I had to interrupt my studies in Berlin. I had gone as far as to talk to Schrödinger about a possible theoretical thesis, but soon we had both decided that the time had come to leave Germany. I wrote to a number of physicists. Rutherford was the first to answer, accepting me at the Cavendish. I came up to Cambridge, I believe in August, to find out more details about the life of a student there. I had heard that the cheapest way to study was to become a member of Fitzwilliam House, but when I met Chadwick and told him about this he made the somewhat cryptic remark: 'If I were you, I would join a college. They do things for you'. Then I met David Schoenberg, and he pointed towards Trinity Street and said 'Well, the good colleges are in this direction: Trinity, St John's and', then he wanted to give me a chance, 'Magdelene'. Magdelene by that time was somewhat known in physics because Blackett had been a fellow there.

Well fortunately the street names changed conveniently with each college so they were easy to find. I walked over to Trinity and was told that they were already full up. At St John's I heard that they would let me know in six weeks. At Magdelene the Senior Tutor, who was then V S Vernon-Jones, said 'Ah, you are a refugee; I suppose we ought to have one'. Then he added: 'I suppose you have no money; we had better give you a hundred pounds.' Now in those days this was about half of what you needed to survive for a year as research student. When I tell this story, people find it impossible to believe there was so little red tape in those days. Maybe one lesson of reminiscences is: can we bring back some of the 'good' parts of the 'good' old days? Well anyhow, as you see, Chadwick had given me very good advice.

Since I wanted to do theoretical work I was assigned to Ralph Fowler. I worked on a little theoretical paper trying to understand the results of Cockcroft and Walton, who had bombarded the two different lithium isotopes with hydrogen and deuterium and found very interesting relations among the cross sections which seemed at first sight puzzling.[1] In particular it was hard to understand why the reaction between lithium-6 and a deuteron would work well, when the lithium-6 spin was given in the tables as zero. The spin of the deuteron was already known as 1, and an S wave would not have produced a state which would go into two alpha-particles. So I ventured to suggest that lithium-6 perhaps had a spin of 1 and a small magnetic moment, so that it had not been seen before. This turned out to be correct and Rabi and Fox proved this very soon. But in the course of writing this paper I had to note the masses of the lithium atoms, and when I asked Mott whom I could consult on this he said that Chadwick knew all these things. So I went to see Chadwick and we had a long discussion in which he gave me the lithium masses as he knew them, and in the course of this I had the courage to suggest to him the photodisintegration of the deuteron, which I had already been thinking about for a year.[2] In those days you did not feel the hot pressure of those competing. You could sit on an idea for a while and if you lost it you could work on another one.

Anyhow, at first Chadwick, in his usual taciturn way, did not show much interest. But then I mentioned that out of this we should get a better mass for the neutron, and there his interest picked up, because he had something of a feud with the Joliots on the one hand, who had postulated a very heavy neutron, and with Ernest Lawrence on the other hand who had proposed a very light one; his own mass was somewhere in between, though lighter than the proton. About six weeks later, I had written a note for *Nature* on something which you would now call delayed neutrons—it was then clearly a premature note on delayed neutrons—and wanted to show it to Chadwick because you would not have sent anything to *Nature* in the Cavendish if you were just a student.[3] It had to go through the Professor's hands, so Chadwick would probably have had to carry it to Rutherford for permission to publish. While I showed it to him he said: 'Were you the one who suggested the photodisintegration to me? Well it works, it worked last night. Would you like to work with me on this?' It was a very generous offer, considering that he had already done the work. I immediately said yes, and then got Fowler's blessing. Since Rutherford was formally in charge of all research students anyway, whether they were doing theoretical or experimental work, I assume that Chadwick discussed this change with him at one of their daily meetings. Chadwick has recorded that he was in the habit of dropping in around 11 AM every morning and discussing the work of the laboratory with Rutherford. We then worked intensely for about two months, and found that the neutron was definitely heavier than the proton, even heavier than the hydrogen atom. I still remember the shock I felt at the realisation that an elementary particle could decay by beta-emission, because by that time I believed in Heisenberg's statement that the neutron is an elementary particle. Of course the astronomers now would say that the anthropic principle tells you that the neutron must be heavier than the hydrogen atom: were it the other way around we would not be here. But at the time I did not know about that principle. I estimated the half-life of the neutron as about thirty minutes using Sargent's empirical energy–lifetime relation. This was of course an overestimate since it used data from complex nuclei.

We started writing a note to *Nature* sometime in the latter part of July.[4] We noticed an interesting conflict between our results for the photodisintegration cross section and the cross section for the inverse reaction which D N Lea had thought he had observed a year earlier, when he shot fast neutrons through polonium, beryllium and paraffin.[5] It was easy to show by

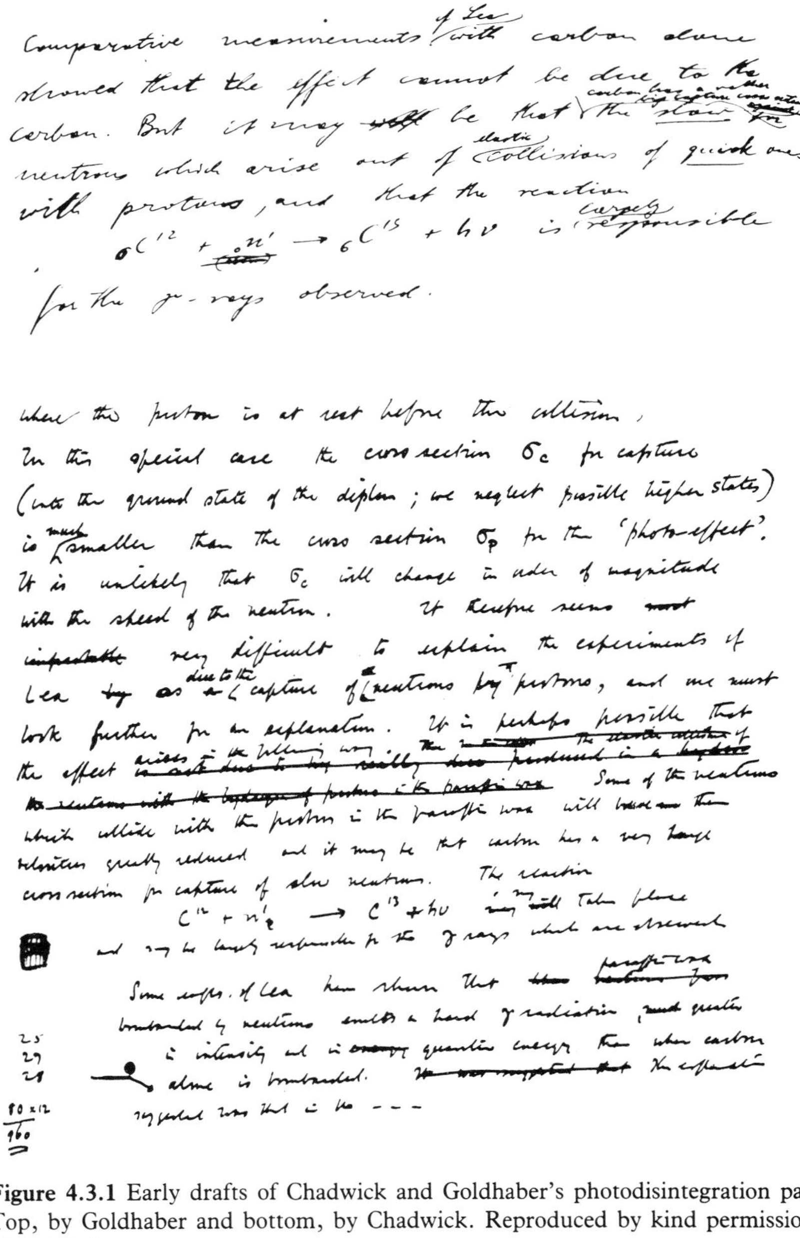

[Handwritten draft, top, by Goldhaber:]

Comparative measurements of Lea with carbon alone showed that the effect cannot be due to the carbon. But it may be that the neutrons which arise out of elastic collisions of quick ones with protons, and that the reaction

$$_6C^{12} + _0n^1 \rightarrow _6C^{13} + h\nu \text{ is largely responsible}$$

for the γ-rays observed.

[Handwritten draft, bottom, by Chadwick:]

where the proton is at rest before the collision. In this special case the cross section σ_c for capture (into the ground state of the diplon; we neglect possible higher states) is much smaller than the cross section σ_p for the 'photo-effect'. It is unlikely that σ_c will change in order of magnitude with the speed of the neutron. It therefore seems very difficult to explain the experiments of Lea as due to the capture of neutrons by protons, and we must look further for an explanation. It is perhaps possible that the effect arises in the following way. Some of the neutrons which collide with the proton in the paraffin wax will have their velocities greatly reduced and it may be that carbon has a very large cross section for capture of slow neutrons. The reaction

$$C^{12} + n^1_0 \rightarrow C^{13} + h\nu$$

will take place and may be largely responsible for the γ rays which are observed.

Some expts. of Lea have shown that the paraffin wax bombarded by neutrons emits a hard γ radiation, much greater in intensity and in quantum energy than when carbon alone is bombarded. The explanation suggested was that it be ---

Figure 4.3.1 Early drafts of Chadwick and Goldhaber's photodisintegration paper. Top, by Goldhaber and bottom, by Chadwick. Reproduced by kind permission of M Goldhaber.

a thermodynamic argument that there was a discrepancy of an order of 1000 or more between our results and Lea's results. Lea's were much too large compared with what we should expect from ours. We therefore speculated that perhaps the neutrons were first slowed down in the paraffin and then captured as slow neutrons, and we included this speculation in early drafts of our paper, which I still have in Chadwick's and my handwriting. After a few days, however, Chadwick said to me: 'Let us not speculate'. By that time I had absorbed enough of the Rutherfordian spirit against speculation to agree. Rutherford very rarely speculated in public. The famous example to the contrary was the neutron, the concept of which was based on one of his rare mistakes. However when a genius makes a mistake it can free his imagination. This is not inviting us all to make mistakes, but we are perhaps a little too hard on people these days when they do. A few months later, when Fermi and his collaborators discovered slow neutrons, the paradox we had noted was resolved. Rutherford sought me out and was uncharacteristically agitated about the fact that this discovery had been made outside the Cavendish. He made some soothing remarks. A psychohistorian of science might suspect that it was he who had originally talked Chadwick out of that speculation at one of their daily meetings but I do not know.[6]

When we continued the work on photodisintegration we wanted to measure photoneutrons coming out from a large amount of heavy water, and at that time Rutherford had a good fraction of the world's supply under his command. I went to him to ask permission to use it, and he wrote a note to Oliphant which I kept. It said: 'Hand over 25cc of heavy water to Goldhaber for the time being'. I hope I returned it.

After the photodisintegration work and Fermi's discovery of slow neutrons we worked on slow neutron reactions in lithium, boron and nitrogen, reactions which have since become of some interest to other scientists and to technical people. The lithium-6 reaction is a way of making tritium, which decays into helium-3, and helium-3 is now widely available for low temperature work: I call it a 'cold-war surplus'. The nitrogen-14 makes carbon-14 which is of course valuable for other things. So we had a lot of interesting results in a very short time, due largely to Chadwick's preparation for this work, a preparation which had gone on for many years. He had modern ionisation chambers and recording equipment with which to record all these disintegrations happening; and he had an incredibly good assistant called Mr Nutt, who helped me in all my work, and who is known from a line in the 'New Cavendish Alphabet', composed and recited by Normal Feather at the Cavendish dinner of 1934: 'N is for Nutt, who discovered the neutron'. Whenever Chadwick was not around Nutt and I would be able to do what we liked. When Chadwick was around, we had to be more cautious.

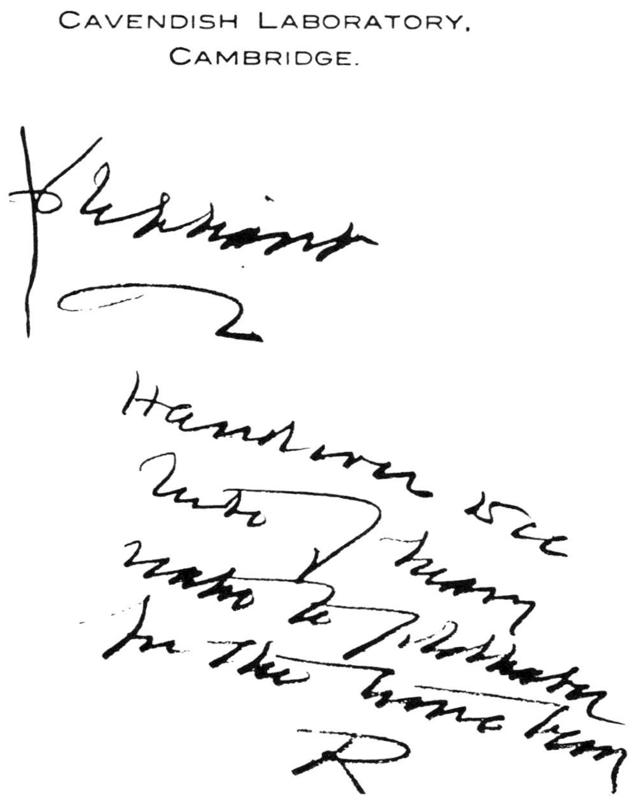

CAVENDISH LABORATORY,
CAMBRIDGE.

Figure 4.3.2 Rutherford's authorisation to Oliphant to lend some heavy water to Goldhaber. Reproduced by kind permission of M Goldhaber.

Notes

1   Goldhaber M 1934 *Proc. Camb. Phil. Soc.* **30** 561
2   See Goldhaber M 1979 The nuclear photoelectric effect and remarks on higher multipole transitions: A personal history in R Stuewer (ed) *Nuclear Physics in Retrospect* (Minneapolis: University of Minnesota Press) p. 81
3   Goldhaber M 1934 *Nature* **134** 25
4   Chadwick J and Goldhaber M 1934 *Nature* **134** 237
5   Lea D N 1934 *Nature* **133** 24
6   See also Goldhaber M 1971 *Proc. R. Soc. Edin.* A **70** 191

4.4   Reminiscences of Cambridge in the Thirties

R Peierls

My first glimpse of Cambridge was in the summer of 1928 when, as a gradute student in Heisenberg's department at Leipzig, I came to England for my summer holiday. I had enrolled in a summer course run by the Extra-Mural Department, which was on literary and cultural subjects. I was fascinated by England, so completely different from the Continental countries I knew, and particularly by Cambridge: the architecture, the student digs in which I stayed, grubby and primitive, with the ground floor windows secured to make students keep the prescribed hours, but presided over with great warmth by the landlady, the unfamiliar bacon-and-egg breakfast, and the tea parties given by academic families for members of the summer course. All this was strange, but it appealed to me, and I felt this would be a good place to come back to.

Near the end of my holiday, when the physicists were back, I did come back to Cambridge, and I called on Dirac, whom I had first met in Leipzig. In fact, I had been delegated one evening to take him to a theatre. In Germany you were not then allowed to take hats or coats into the auditorium, but had to leave them in the cloakroom. It was summer, and I had no coat or hat, but Dirac had a hat, which I tried to persuade him to deposit. But he refused, and I remember worrying about the embarrassment that might result if his breach of the rules was noticed. Nothing happened of course.

So I presumed on our acquaintance to look him up in Cambridge. He was very charming and arranged for me to use the Cavendish Laboratory, and introduced me to R H Fowler. When Fowler heard that I came from Leipzig, he was delighted. 'No doubt you can tell us about the new work by Bloch. We have no speaker for the next meeting of the Kapitza Club, so you must talk there'. Neither my English nor my familiarity with the details of Bloch's paper were really adequate, but it never occurred to me to refuse. I do not remember much of the meeting, but at least there was no disaster.

Otherwise my main recollection is of trying to read in the unheated and bitterly cold laboratory library (it was by now late September). On one

195

occasion I overheard one young man, who was studying a German paper, ask another whether the word 'Blei' might perhaps mean lead, and I was able to confirm his conjecture, with the satisfaction of having made some very small contribution to the progress of physics at the Cavendish.

I came back to the Cavendish in very different circumstances in April 1933. I had a one year Fellowship from the Rockefeller Foundation, of which I had spent the first half with Fermi in Rome, and had arranged to be in Cambridge for the summer, in contact with Fowler and Dirac. This was the beginning of the Hitler era in Germany, and my wife and I were not intending to go back to Germany. After the end of the Fellowship in October our plans were not clear, and our first child was expected the following August.

When we had found somewhere to live for a while, I set out to call on Fowler in the Cavendish. I went up to the first floor, where the offices were, and found a row of unmarked doors, and nobody about to ask. After pacing up and down for a while, I decided to try the most inconspicuous looking door to ask for instructions. I knocked and entered what turned out to be Rutherford's room, but only his secretary was there, who told me where to find Fowler.

Fowler was an old acquaintance from the previous visit, as was Dirac. I had met Blackett in Zurich, and the Blackett's took us under their wing and helped us settle in Cambridge. I met the rest of the physicists before long. My recollection of these encounters tends to fuse the summer of 1933 with the period of 1935–7, which I also spent in Cambridge, and even with brief visits during the intervening years, so I shall not attempt to distinguish.

Of course we soon got to know Rutherford, both at meetings in the Cavendish and at some of the parties held regularly at his house. It goes without saying that I found his command of physics and his energy as impressive as his reputation had led me to expect. What one had not been led to expect was his simple directness and his warm interest in people. He loved to tell stories, and I recollect, for example, one of his favourite stories about the new university library being opened by King George V and Queen Mary. An enthusiastic librarian showed the royal visitors the bookshelves, which were of a new design, and had certain advantages. To show interest the King asked what they were made of. 'Of steel, your Majesty'. 'Will they last?' This earned a poke in the ribs from the Queen's umbrella: 'Don't be silly, George; of course they will last!' Another of his reminiscences was of the earlier days in Cambridge, when the social life was strictly stratified, so that professors would tend to visit each other, and the heads of colleges would keep to themselves; the wife of a master was heard to remark 'I wish I knew what they talk about in a professor's house'.

It is said that Rutherford did not think much of theoreticians, but I never experienced any lack of consideration or courtesy. When I came back to Cambridge in 1935 to an appointment at the Mond Laboratory, of which

Rutherford had taken charge when Kapitza was unable to return from Moscow, I met him in the courtyard a few days after my arrival. I was not sure he would remember who I was, but he greeted me and asked how I was settling down in Cambridge. 'Have they paid you any money yet?' I said no, that would come at the end of the month. 'Will you be able to last until then?' I know many a professor with less on his mind, who would not remember the practical problems of young people. This conversation with Rutherford set me an example which, I hope, I have always remembered when necessary.

I remember one occasion which showed his capacity for introducing some badly needed informality. He was visiting Manchester to give a colloquium talk, and he was introduced by Lawrence Bragg, who spoke at length about his feeling of inadequacy in succeeding Rutherford as Professor at Manchester. People on the Faculty Board, he said, had been accustomed to turn for wisdom on difficult decisions to the Professor of Physics, and were disappointed to see it was only him sitting there. This left the audience with some feeling of embarrassment, which was happily defused when Rutherford got up and said 'Professor Bragg has expressed some doubts about his ability to fill my chair, but' and here he pointed to his own bulk and to the beginning of middle-age spread of Bragg, 'I think he is well on the way to doing so'.

Initially my closest contacts were with the theoreticians; with Fowler and Dirac, and the younger ones, Nevill Mott, H R Hulme and H M Taylor.

Dirac had the reputation of being rather silent, but this is not an accurate description—he does talk, but not in idle conversation, only when he has something to say. What he does say is often unexpected, because his reaction to an idea is never trivial, and in this respect his style in conversation is like that in his physics. Once at tea in our house the conversation was about the fact that all recent children born to physicists in Cambridge had been girls, and someone remarked 'It must be something in the air'. After a pause, Dirac added 'Or perhaps in the water'.

He always answers direct questions, but does not feel the need to enlarge upon them. On questions about physics he is likely to answer yes or no, if he has thought out the problem, or 'I don't know' otherwise. This does not make it easy to engage him in debate on unsolved questions. The same tendency can cause misunderstandings, as on one occasion when both Wigner and I (then in Manchester) were visiting Cambridge and separately called on Dirac to see the experiment on isotope separation in which he was then engaged. One could see gas from a compressor entering a little brass T-piece, with hoses coming out on both sides. Eventually light gas was to emerge on one side and heavier gas on the other. One could feel that the gas on one side was hot and that on the other was cold, so evidently something interesting was going on. Back in Manchester, Wigner complained that Dirac was secretive and refused to say what the principle of his device was.

This had not been my experience, so I probed a little more, to discover that all Wigner had said was 'It must have been very difficult to make the little brass piece' to which Dirac replied 'No, that was quite easy'; He had been asked a simple question and he had answered it, whereas Wigner, in Hungarian style, had asked for information, and had been refused. Wigner would not accept my interpretation so I bet him that, if he had asked Dirac for an explanation, it would have been given. When I later asked Dirac if he would have explained on demand, he said 'I do not know'. He never liked hypothetical questions.

Nineteen thirty three was a time of economic depression, and academic positions were in short supply. It would not have been surprising if the Cambridge physicists, and particularly the theoreticians, had resented the arrival of so many refugees, who would increase the competition for posts. Instead the new arrivals were received with great kindness, and physicists took an active part in organising the Academic Assistance Council, which became later the Society for the Protection of Science and Learning, to give help and support to refugee scientists until they could find their feet. I remember one manifestation of this spirit: I had applied for an Assistant Lectureship at Manchester, and an informal reply from Bragg had been encouraging. Then one day a letter for me arrived from Manchester. Nevill Mott, who knew of my situation, saw the letter in the Cavendish, and cycled to my house in great excitement to bring it to me. (The letter turned out, in fact, to be a rejection, but later I was offered a grant from local Manchester funds similar to the Academic Assistance Council.)

The same year was also an exciting time in nuclear physics. The neutron was newly discovered. Cockcroft and Walton's accelerator was new and others, particularly Oliphant and Dee, were starting on the technique of using artificially accelerated particles. Chadwick, the discoverer of the neutron, was a particularly good person to learn from, once one had got used to his manner. One day, I had come back from Manchester for a visit to learn what was new in physics, and, coming to the Cavendish, I saw Chadwick. He greeted me and asked how long I had come back for. I explained and said if he had some time to spare I would also like to have a talk with him. He looked at me over the top of his glasses and said 'Yes? What for?' Luckily I had seen enough of him not to be discouraged; I explained what I wanted, and we had a useful talk.

About that time Chadwick was engaged in observing the photodisintegration of the deuteron. He had not told anyone of this experiment yet but during one visit he teased Bethe and me by saying 'I bet you could not calculate the photodisintegration cross section of the deuteron!' We thought we could, and this started us off in nuclear physics. Chadwick's work on this was done in collaboration with Maurice Goldhaber, another refugee, then very young and not very experienced in the ways of the world. He caused some raised eyebrows by going around telling everybody what experiments they ought to be doing and how—what made this particularly irritating was that he was usually right.

When I came back to Cambridge in 1935 I had a research appointment in the Mond Laboratory. Since Kapitza was detained in Moscow and could not resume his Royal Society Professorship, it was decided to use the money thus saved to establish two Research Fellowships, to which J D Allen and I were appointed. The Mond Laboratory had been built fairly recently for Kapitza's magnetic and low-temperature work and was in the centre of the courtyard containing the Cavendish and some other laboratories, with access through gates from Free School Lane and Downing Street, which were firmly locked at a certain hour in the evening. To keep them open beyond this hour was as unthinkable as for anyone, however senior, to get a key to them. Indeed when Chadwick had once to continue some measurements until late at night, he solved the problem by bringing a camp bed into the laboratory and spending the night there.

After I had been there for a while, it was suggested that I might give some regular lectures. This was a little difficult because to lecture one had to have a Cambridge MA. But the experts soon found a solution. It involved passing three resolutions, or 'Graces' in the Senate House; one to create a new unpaid position of Assistant-in-Research at the Mond Laboratory, the second to add this post to the list of positions whose holders could be awarded an ex officio MA, and the third to appoint me to that position and to give me the MA.

Another marginal contact with Cambridge teaching came in my second year when I was asked to supervise three final-year undergraduates for St John's College. I had no idea what a supervisor was supposed to do, and nobody seemed to be able to give me a clear explanation, so I don't know what relation my conversations with the students had to the Cambridge tradition. Still, it seems that no irreparable harm was done. One of the three was Charles Kittel, now at Berkeley, another, Rivlin, later became well known in polymer research, and the third, Huck, went into schoolteaching to earn some money because his father died when he was about to graduate. He is now on the staff of the University of Surrey.

This teaching was done for St John's College, and I was given limited dining rights ('up to once a week in term time at your own expense'). The Fellows I met there included Cockcroft, and this was a good opportunity to see him in a relaxed mood and willing to chat in a leisurely way. Otherwise he was very busy, looking after nuclear experiments and the building of the new High-Voltage Laboratory, as well as the Mond Laboratory in the absence of a director. He also dealt with maintenance and restoration of College buildings. Driving around the country he would often notice a dilapidated wall of old bricks; he would then stop and buy the bricks, to replace modern machine-made bricks which had been used for repairs in the college. With all the activities he had to make very efficient use of his time, and could not spare any for idle converastion, so the encounters in the combination room after dinner were exceptional.

One episode which made a deep impression on me took place in the Mond Laboratory, where I worked in a room which also contained a drawing board. Cockcroft was busy with a drawing when the secretary came in to say the architect was on the phone and wanted to know whether the light switches in the New High-Tension Laboratory were to be black or brown. (In those days switches were either black or brown.) Anybody else would have taken some time for reflection and would have said 'It does not really matter, but, let me see, perhaps brown. . .'. Cockcroft did not lift his eyes from the drawing, and without a moment's hesitation said 'Brown' (or perhaps it was black, I don't remember).

One institution I remember from the early 1930s was the Physics Club. This was a colloquium held at irregular intervals (once or twice a term I think) started on the initiative of exiles from the Cavendish who felt isolated in London or elsewhere, and wanted to keep in touch. The meetings were usually held in the Cavendish or in London. These meetings were greeted with great enthusiasm by the German refugees, who were eager to participate. However, membership of the club was limited; many of the 'old hands' like G P Thomson and Darwin did not want the meetings too large, and said that in a semi-public meeting they would be reluctant to ask stupid questions. The outsiders—including myself—could not understand this attitude, and thought it was unreasonable. I still thought so after I had been admitted to membership of the club. However, at first the rule was respected. But it was not, of course, possible to enforce it: one could not have a steward at the door asking for people's credentials, and gradually many non-members invaded the meetings. Perhaps this caused the decline of the club, but I do not remember for how long it continued—perhaps I just lost touch gradually after leaving Cambridge.

In conclusion I would like to stress that this collection of anecdotes, most of them light-hearted, should not obscure the feeling of excitement with which I recall the period I was privileged to spend in one of the world's centres of physics, combining its ancient tradition of institutions with revolutionary new methods and ideas in physics.

Select Bibliography

1. Archival material

The main archival source for the history of twentieth century physics is the microfilm archive, *Sources for the History of Quantum Physics*. Drawn from collections around the world, this is to be found in the following locations: American Philosophical Society, Philadelphia, Pennsylvania; University of California at Berkeley; American Institute of Physics, New York; Niels Bohr Institute, Copenhagen; Science Museum, London (forthcoming). A partial catalogue has been published: T S Kuhn *et al* 1967 *Sources for the History of Quantum Physics* (Philadelphia: American Philosophical Society). The archive is cited in the footnotes as *SHQP*.

Other collections relevant to the subject matter of this volume, some of which are wholly or partially reproduced in *SHQP*, include the following:

Niels Bohr Archive, Niels Bohr Institute, Copenhagen: correspondence between Bohr and Blackett, Chadwick, Cockcroft, Dirac, Fowler and Rutherford.

Churchill College Archives, Cambridge: papers and correspondence of Chadwick, Cockcroft, Dirac, Walton and C T R Wilson.

Trinity College Library, Cambridge: papers of G I Taylor.

Cambridge University Library: papers and correspondence of Rutherford.

University of Texas at Austin: correspondence from Rutherford *et al* to O W Richardson.

Bodleian Library, Oxford: correspondence from Blackett to Peierls.

Royal Society, London: papers and correspondence of Blackett.

A complete catalogue of correspondence between physicists during the twentieth century is being prepared by the Office for History of Science and Technology, University of California at Berkeley, and should become available shortly.

2. Royal Society obituaries

These constitute the most important source of biographical information on deceased British scientists. For convenience, they are listed here by subject rather than author. The letters *ON* or *BM* refer to the *Obituary Notices of Fellows of the Royal Society* and to the *Biographical Memoirs of Fellows of the Royal Society* respectively, the latter having replaced the former title. As with the other sections of this bibliography, a selection of the more important papers only is given.

F W Aston: 1945–8 *ON* **5** 635–51
J D Bernal: 1980 *BM* **26** 17–84
P M S Blackett: 1975 *BM* **21** 1–115
J Chadwick: 1976 *BM* **22** 11–70
J D Cockcroft: 1968 *BM* **14** 139–88
C D Ellis: 1981 *BM* **27** 199–234
N Feather: 1981 *BM* **27** 255–82
R H Fowler: 1945–8 *ON* **5** 61–78
L H Gray: 1966 *BM* **11** 195–218
J L Pawsey: 1964 *BM* **10** 229–44
G I Taylor: 1976 *BM* **22** 565–633
C T R Wilson: 1960 *BM* **6** 269–95

3 Historical works and recollections

(*a*) Collections

Birks J B (ed) 1962 *Rutherford at Manchester* (London: Heywood). This includes valuable papers by Blackett and Bohr, the latter also published in *Proc. Phys. Soc.* **78** 1083–115 (1961)
Nobel Lectures in Physics, 1922–1941 1965 (Amsterdam: Elsevier)
Nobel Lectures in Physics, 1942–1962 1964 (Amsterdam: Elsevier)
Notes and Records of the Royal Society vol 27 1972 pp 1–94 Rutherford centennial celebrations. This volume contains relevant papers by Blackett, Feather, Lewis, Massey, Mott and Oliphant.
Proc. R. Soc. A **371** 1–177 (1980) The beginnings of solid state physics. A symposium. This includes papers on the early 1930s by Bethe, Peierls and A H Wilson.
Proceedings of the Tenth International Congress of the History of Science, 1962 1964 (Paris: Hermann) pp 121–62. This includes papers by Chadwick, Feather, Purcell and Segré.
Stuewer R (ed) 1979 *Nuclear Physics in Retrospect. Proceedings of a Symposium on the 1930s* (Minneapolis: University of Minnesota Press). This includes valuable papers by Goldhaber and McMillan and also forms the best existing introduction to the early history of theoretical nuclear physics generally.

Weiner C and Hart E (eds) 1972 *Exploring the History of Nuclear Physics* (New York: American Institute of Physics)

(*b*) Miscellaneous books and articles

Allibone T E 1973 *Rutherford: The Father of Nuclear Energy* (Manchester: Manchester University Press)
Biew A M 1956 *Kapitsa* (London: Frederick Muller)
Blackett P M S 1954 The birth of nuclear science *The Listener* **51** 380–2, 424–5, 477–8
—— 1969 The early days of the Cavendish *Rev. Nuovo Cimento* xxxii–xl
Bromberg J 1971 The impact of the neutron on Bohr and Heisenberg *Hist. Stud. Phys. Sci.* **3** 307–41
—— 1976 The concept of particle creation before and after quantum mechanics *Hist. Stud. Phys. Sci.* **7** 161–91
Brown L M 1978 The idea of the neutrino *Phys. Today* **31** (September) 23–8
Cassidy D 1981 Cosmic ray showers, high energy physics and quantum field theory: programmatic interactions in the 1930s *Hist. Stud. Phys. Sci.* **12** 1–39
Chadwick J 1972 Search for the neutron *Adv. Exp. Phys.* **1** 193–7
Cockcroft J D 1946 Rutherford: life and work after the year 1919, with personal reminiscences of the Cambridge period *Proc. R. Soc.* **58** 625–33. (Reprinted in *Rutherford by Those who Knew Him* (London: Institute of Physics, 1954))
Condon E U 1978 Tunneling—how it all started *Am. J. Phys.* **46** 319–23
Crowther J G 1974 *The Cavendish Laboratory 1874–1974* (London: Macmillan)
Devons S 1971 Recollections of Rutherford and the Cavendish *Phys. Today* (December) 39–45
Douglas A V 1956 *The Life of Arthur Stanley Eddington* (London: Nelson)
Evans I B N 1940 *Man of Power: The Life Story of Baron Rutherford of Nelson, OM, FRS* (London: Penguin)
Eve A S 1939 *Rutherford: Being the Life and Letters of the Rt. Hon. Lord Rutherford, OM* (Cambridge: Cambridge University Press) This is the standard biography.
Feather N 1940 (reprinted 1973) *Lord Rutherford* (London: Priory Press)
—— 1960 A history of neutrons and nuclei *Contemp. Phys.* **1** 191–203, 257–66
—— 1960 Rutherford's Cavendish *New Sci.* **7** 598–600
Gamow G 1972 *My World Line. An Informal Autobiography* (New York: Viking Press)
Goldhaber M 1971 Remarks on the prehistory of the discovery of slow neutrons *Proc. R. Soc. Edin.* A **70** 191–5

Goldstein S 1969 Fluid mechanics in the first half of this century *Ann. Rev. Fluid Mech.* **1** 1–28

Hall A R 1969 *The Cambridge Philosophical Society: A History, 1819–1969* (Cambridge: Cambridge Philosophical Society)

Hanson N R 1963 *The Concept of the Positron* (Cambridge: Cambridge University Press)

Kapitza P L 1966 Recollections of Lord Rutherford *Proc. R. Soc.* A **294** 123–37 (reprinted in Kapitza P L 1967 *Collected Papers* vol 3 (Oxford: Pergamon))

—— 1967 Recollections of Professor E Rutherford in *Collected Papers* vol 3 (Oxford: Pergamon) pp 22–37

—— 1971 Rutherford and creativity in science *New Sci.* **51** 639–40

Kay W A 1963 Recollections of Rutherford. Being the personal reminiscences of Lord Rutherford's laboratory assistant published for the first time *Nat. Phil.* **1** 129–55

Kevles D J 1972 Towards the annus mirabilis: nuclear physics before 1932 *Phys. Teacher* **10** 175–81

Kragh H 1979 Methodology and philosophy of science in Paul Dirac's physics *IMFUFA Text 27* (Roskilde University Centre)

Larsen E 1962 *The Cavendish Laboratory: Nursery of Genius* (London: Edmund Ward)

Lewis W B 1971 Frontier events in electrical, electronic and radio engineering from picowatts to terrawatts in the last 40 and next 60 years *IEE Centenary Lecture* (Chalk River, Ontario: Atomic Energy of Canada Limited)

—— 1979 Early detectors and counters *Nucl. Instrum. Meth.* **162** 9–14

Mann F G 1976 *Lord Rutherford on the Golf Course* (Cambridge: Heffer)

Massey H W 1977 Patrick Blackett: an appreciation *Proc. Ind. Nat. Sci. Acad.* A **43** 1–17

Mehra J 1972 The golden age of theoretical physics: P A M Dirac's work from 1924 to 1933 in A Salam and E P Wigner *Aspects of Quantum Theory* (Cambridge: Cambridge University Press)

Moon P B 1974 *Ernest Rutherford and the Atom* (London: Priory Press)

Neményi P 1962–6 The main aspects and ideas of fluid dynamics in their historical development *Arch. Hist. Exact Sci.* **2** 52–86

Niblett C A 1980 Images of progress. Three episodes in the development of research policy in the UK electrical engineering industry *PhD Thesis* (University of Manchester)

Oliphant M L E 1972 *Rutherford: Recollections of the Cambridge Days* (Amsterdam: Elsevier). This is the best single account of the Cavendish in the early 1930s.

Pollard E C 1968 *Physics: An Introduction* (Oxford: Oxford University Press) ch 1

Ratcliffe J A 1978 Wireless and the upper atmosphere, 1900–1935 *Contemp. Phys.* **19** 495–504

Rossi B 1981 Early days in cosmic rays *Phys. Today* **34** (October) 34–41

Sargent B W undated *Recollections of the Cavendish Laboratory Directed by Rutherford* (Queens University, Kingston, Ontario, Nuclear Physics Laboratory)

Satterly J 1939 The post-prandial proceedings of the Cavendish Society *Am. J. Phys.* **7** 179–84, 244–8

Shire E S 1972 *Rutherford and the Nuclear Atom* (London: Longman)

Snow C P 1958 The age of Rutherford *Atlantic Monthly* **202** (November) 76–81

Stratton F J M 1949 *The History of the Cambridge Observatory* (Cambridge: Cambridge University Press)

Swann W F G 1955 *The Story of Cosmic Rays* (Cambridge, Mass.: Sky Publishing)

Thomson G P 1964 *J J Thomson and the Cavendish Laboratory in His Day* (London: Nelson)

Walton E T S 1982 Recollections of nuclear physics in the early nineteen thirties *Europhys. News* **13** (August) 1–3

Weiner C 1972   1932—moving into the new physics *Phys. Today* **25** (May) 40–9

—— 1974 Institutional settings for scientific changes: episodes from the history of nuclear physics, in A Thackray and E Mendelsohn *Science and Values* (New York: Humanities Press)

Wilson C T R 1960 Reminiscences of my early years *Notes and Records of the Royal Society* **14** 163–73

Name Index